Keyla M. Alencar da Silva Alves
Ranyére Nobrega

Socio-environmental vulnerabilities in the Brazilian semi-arid region

Keyla M. Alencar da Silva Alves
Ranyére Nobrega

Socio-environmental vulnerabilities in the Brazilian semi-arid region

A study on exposure, sensitivity and adaptive capacities to climate change

ScienciaScripts

Cover image: www.ingimage.com

This book is a translation from the original published under ISBN 978-3-330-76778-2.

Publisher:
Sciencia Scripts
is a trademark of
Dodo Books Indian Ocean Ltd. and OmniScriptum S.R.L publishing group

120 High Road, East Finchley, London, N2 9ED, United Kingdom
Str. Armeneasca 28/1, office 1, Chisinau MD-2012, Republic of Moldova, Europe
Managing Directors: Ieva Konstantinova, Victoria Ursu
info@omniscriptum.com

Printed at: see last page
ISBN: 978-620-8-57554-0

CONTENTS

CHAPTER 1

SOCIO-ENVIRONMENTAL VULNERABILITY, CLIMATE CHANGE AND OTHER CONCEPTS

Climate change can be seen as a risk management problem, where changes in climatic factors (mainly temperature and precipitation) are the result of anthropogenic interventions, while vulnerability arises from the development model and the use of natural resources. To reduce the magnitude of the risk, it is proposed to reduce unsustainable human activities that produce impacts, this action is defined as mitigation. To reduce vulnerability, Adger (2006) proposes developing and applying adaptation strategies to climate change. However, in order to suggest adaptation measures it is necessary to understand and assess vulnerability to climate anomalies.

Vulnerability is made up of the physical, social, economic and environmental contexts of a given region, sector or social group susceptible to the impacts of climatic phenomena. The dynamics of vulnerability, as a multifactorial element, must have data that makes it possible to analyse the recent past and project it into the future in order to analyse the possible impacts of climate change. Quantitatively and qualitatively assessing vulnerability means spatially prioritising where adaptation measures are most needed.

Why analyse vulnerability? This question should be challenged by recognising it as a key element in stimulating the potential impacts of climate change. By quantifying the data, it is possible to define how vulnerable a particular social group or locality is, eliminating ambiguities about what or who is more or less vulnerable.

The aim of this chapter was to develop a conceptual proposal to contribute to the construction of the climate change vulnerability index, and for this proposal to be made up of concepts and indicators.

It is necessary to have detailed knowledge of the indicators that make up the index, understanding their dynamics in such a way as to be able to build a conceptual model of the relationship between the climate and the indicators on the scales of time and space. With the conceptual model and the index built, it is possible to diagnose the causes of vulnerability and the factors that make it possible to characterise near future scenarios.

Creating a vulnerability index using indicators is the centrepiece of this research . Applying it requires estimating the recent risk and comparing it, on a spatio-temporal scale, to the impacts when they present significant climate threats.

Therefore, before describing the technical procedures used to create the index, it is important to conceptualise the terms used here. This chapter will therefore consist of three stages: a) conceptualisation of the terms (conceptual model); b) selection and definition of the indicators; c) technical procedures for creating the climate change vulnerability index.

1.1. Risk and environmental impact

In general terms, risk is understood as the probability that an event - expected or not - will come true (DAGNINO and CAPRI, 2007). Considering probability as the operating mechanism of risk, the definition of the concept states that risk presents itself in situations or areas where there is susceptibility, vulnerability and exposure to the occurrence of some kind of threat, danger, problem, impact or disaster (AMARO, 2005).

Environmental risks are understood to be those risk situations that are not isolated from the actions of their surroundings - the environment in its broadest sense - whether the environment is natural or man-made (social and technological). Veyret and Meschine de Richemond (2007, p.63) conceptualise environmental risks as: "resulting from the association between natural risks and risks arising from natural processes aggravated by human activity and occupation of the territory".

However, it is important to emphasise that despite the concepts and definitions of risk applied as indicators of environmental problems, risk is a social object that, directly or indirectly, refers to human beings as individuals or social groups. As Veyret (2007, p. 11) states:

> There is no risk without a population [social being] or individual [biological being] that perceives it and could suffer its effects. Risks are taken, assumed, refused, encouraged, evaluated and calculated. Risk is the translation of a threat, a danger for those who are subject to it and perceive it as such.

Impacts are considered to be the "materialisation of risk", which means that in some cases impacts can reach critical levels. Explaining the impact requires documenting the characteristics of risk, danger and vulnerability, not just considering the physical approach that explains the disaster solely as an expression of the forces of nature (CASTRO, 2003).

The cross-referencing of information that characterises risk and vulnerability is generally used to build the first scenarios of impacts caused by climate change, defining an image of the future essentially based on dangers, without considering the dynamics of vulnerability. It is now recognised that vulnerability is the fundamental element for understanding impacts, since systems and their respective functioning depend on many more elements than just the climate (PARMESAN, 2006).

1.2. Vulnerability

Vulnerability represents the interface between exposure to physical threats to human well-being and the capacity of people and communities to control such threats. Threats can arise from a combination of physical and social processes. Vulnerability is a variable indicator, so it must be measured using quantitative and qualitative methods. However, reducing vulnerability implies detecting which factors contribute to its increase and identifying the points where you can intervene in the network of causes of hazards and their human consequences (CUTTER, 2011).

According to the fourth IPCC report (2012, p. 87) "vulnerability is the degree to which a system is susceptible to and unable to cope with the adverse effects of climate change, including climate variability and extreme events". Vulnerability must be analysed as a systemic, multifactorial, multisectoral, multitemporal and multiscalar condition, and like danger, vulnerability is dynamic.

Exposure plus sensitivity are the factors that generate vulnerability, but exposure makes a greater contribution to vulnerability, so that if there is no exposure to a specific phenomenon, even if there is sensitivity, there will be no risk and therefore no vulnerability. Risk is the combination of danger and vulnerability, so a given system is vulnerable to the extent that it is exposed to danger (Figure 1).

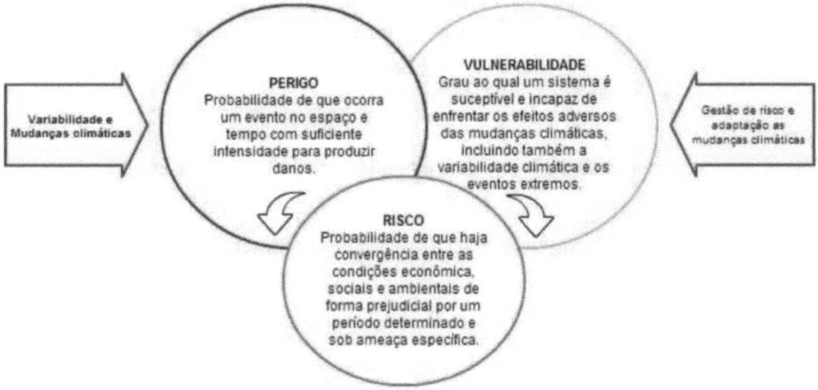

Figura 1: Diagram of the relationship between danger, risk and vulnerability.
Source: CUTTER, 2011.

To measure the vulnerability of a particular social group or area, it is common to use indicators related to physical, social and economic factors. The relationship between indicators such as population density (social indicator) and Gross Domestic Product (GDP) (economic indicator) can provide results that characterise people's vulnerability to extreme weather conditions. The indicators, converted into indices, lead to the quantification of vulnerability and its dynamics.

1.3. Elements for assessing vulnerability

Vulnerability is characterised by the exposure, sensitivity and adaptive capacity of each system under the influence of climate variability, extreme events and climate change (MONTERROSO, 2012). Vulnerability, in general terms, consists of a set of conditions of systems[1] that derive from the prevailing historical and cultural, social, environmental, political and economic contexts:

> Vulnerability is the set of physical, social and economic conditions that affect people in a social or natural system, due to the occurrence of natural phenomena.
> (CUTTER, 2003, p 243)

Analysing climate change also requires estimating danger and vulnerability and, like any projection into the future, the results will be associated with a degree of uncertainty. Estimating vulnerability is always complex, but the magnitude of the uncertainty is expected to be less than the actual signals presented by the results. Estimating the degree of uncertainty in vulnerability scenarios is difficult and therefore little discussed, because future vulnerability depends on human decisions that are difficult to define. A range of possible future vulnerability conditions and, consequently, uncertainties can be generated.

To estimate vulnerability to climate change, CUTTER (2003) suggests that:

> There is no universally accepted methodology for quantifying future vulnerability which, combined with climate projections, makes it possible to estimate the risk that will be faced under the effects of climate change.
>
> (CUTTER, 2003, p.243)

[1] In this case, he understands "system" as a holistic structure that brings together social, cultural, physical and economic elements (BUCKLEY, 1973).

The concept of vulnerability has gained prominence in studies analysing the potential impacts of climate change, becoming an essential element in the construction of public policies to mitigate damage and impacts.

The approach to the problem of establishing vulnerability must go through three basic questions: 1) who or what is vulnerable? 2) vulnerable to what? 3) why is it vulnerable? (Figure 2).

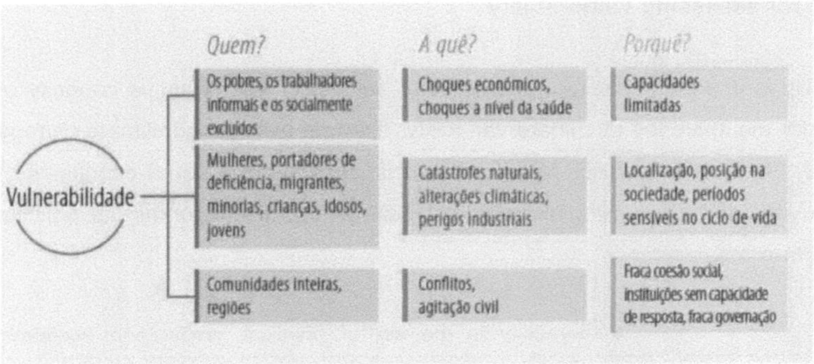

Figura 2: Who is vulnerable to what and why?

Source: Human Development Report (2014).

a. "Who or what is vulnerable? The answer lies in the analysis of the conditions of sensitivity presented by environmental systems and social groups from the characterisation of the object of analysis, as well as in the relations of the object of analysis with danger.

b. "Vulnerable to what? The answer to this question lies in recognising the dynamics of hazard factors and their geospatial relationship with people, making it possible to estimate the degree of exposure in the study area.

c. "Why is it vulnerable? This is the key to diagnosing vulnerability, as it involves analysing the factors that affect systems and recognising their adaptive capacities.

In order to achieve results that contribute to reducing vulnerability, it is necessary to transcend qualitative analysis to a quantitative one that allows the aspirations of sustainable development to be pursued. The qualitative approach is inductive and is based on the expansion of data and information with interpretative means of investigation that start from observations that are not always structured, open interviews, document reviews, group discussions and evaluations of experiences. It can be said that qualitative analyses of vulnerable conditions are not always consistent, especially when developed from different points of view, which would consequently present contrasting results of reality (CUTTER, 2011).

The quantitative approach is deductive, aims to limit information and is linked to structured experiments, interviews with closed questions and standardised measuring instruments. Establishing the conditions for developing vulnerability analyses with a quantitative approach makes it possible to legitimise the information obtained, unify the measurement criteria and objectively qualify the conditions of vulnerability to climate change. This type of approach is the most appropriate for analysing the dynamics of vulnerability and for constructing indices.

CHAPTER 2

SCIENTIFIC METHODOLOGY AND TECHNICAL PROCEDURES

2.1. CHARACTERISATION OF THE STUDY AREA 2.1.1 Object of study

The object of study of this research is the Moxotó Hydrographic Basin. It will analyse environmental vulnerability and the impact it has on the social groups that occupy the basin's territory and that are also under the influence of socio-economic activities that generate impacts and foster vulnerability.

The aim of this research is to quantify and analyse the environmental and social vulnerability of the Moxotó river basin by developing indices with the aim of contributing to the implementation of actions to adapt to climate change. This objective is in line with the central lines of the Bali Action Plan (2007) on adaptation to climate change, which highlight the application of adaptation measures by analysing the vulnerability and response capacity of the landscape. This research also carried out a data survey of the basin to identify the impacts. Climate data and data extracted from satellite images were analysed from a space-time perspective, with the aim of creating a preliminary diagnosis of the level of impacts present in the basin's landscape.

2.1.2. Criteria for selecting the study area

The first criterion was that the study area should be located in the semi-arid region of north-eastern Brazil. Since the aim of the research is to quantify and analyse vulnerability to climate change, it was decided to select an area that was within the zone considered critical and highly susceptible to the impacts of climate change. According to the IPCC's 5th report (2014) all the models

show that the semi-arid region will experience warming by 2100, ranging from 1.5°C - 5.5°C, a slight increase in rainfall and an increase in climatic extremes; significant loss of biodiversity, potential loss of vegetation cover with a consequent increase in exposed soil areas and serious impacts on agricultural activity.

The second criterion for choosing the study area was the spatial cut-out of the river basin, since the research is based on analysing possible changes in climate factors (precipitation and temperature), it would be necessary to analyse the spatial distribution of rainwater and, therefore, the spatial cut-out of the river basin is the most appropriate for this type of analysis. The atmosphere does not have defined spatial boundaries, which makes it difficult to determine and quantify changes without adopting a physical boundary or a specific climatic factor. The cut-off of the river basin favours the analysis because it is delimited by water dividers which, in this case, are important for observing which real water loads from precipitation supply this area. The municipality or micro-region are of a political-administrative nature and are not part of this study.

The concept of watershed adopted in this research is that proposed by Lima and Zakia (2000), who define a "watershed" as a group of lands drained by a river and its tributaries, formed in the highest areas of the relief by water dividers, where rainwater either runs off the surface forming streams and rivers or infiltrates into the ground, giving rise to aquifers. Furthermore, they add a systemic approach to the concept: for these authors, river basins are open systems that receive energy through climatic agents and lose energy through runoff, and can be described in terms of interdependent variables that oscillate around a pattern and, therefore, even when disturbed by anthropogenic actions, are in dynamic equilibrium. Thus, any change in the receipt or release of energy or change in the shape of the system will result in a compensatory change that tends to minimise the effect of the change and restore the state of dynamic equilibrium.

2.1.3. Spatial cross-section

The Moxotó river basin, physiographically inserted entirely within the limits of the northeastern semi-arid region, is located almost entirely in the central portion of the state of Pernambuco and in the western part of the state of Alagoas, covering an area of approximately 9,752.71 km^2 (8,778.05 km^2 of which in the state of Pernambuco and 974.66 km$^{(2)}$) in the state of Alagoas).

The corresponding climate of the basin, according to the Kõppen generic classification, is BShW', which corresponds to the hot and dry type, with average annual rainfall ranging from 300 mm to 900 mm, with rainfall concentrated in the months of January to May. The average annual temperature, relative humidity and potential evaporation are 25 °C, 55% and 3,008 mm, respectively (PERRAZZO, 2002).

However, it is important to emphasise that the climate of the semi-arid region should not be understood solely by the information described in its classification, since it is characterised by its extreme variability and the influence of extreme climatic events, such as the El nino/La Nina. The region has considerable rainfall figures, but the problem lies in the temporal distribution of this rainfall, which in the region is concentrated during the months of May to August, thus configuring the rainy season, with drought in all the other months.

The Moxotó river basin partially covers 15 municipalities, 11 of which belong to the state of Pernambuco, namely Arcoverde, Buíque, Custódia, Ibimirim, Iguaraci, Inajá, Jatobá, Manari, Tupanatinga, Sertânia and Tacaratú, Iguaraci being the municipality with the smallest territorial area within the basin, and 4 in the state of Alagoas, Água Branca, Delmiro Gouveia, Mata Grande and Pariconha (CPRM, 2004) (Figure 3).

The Secretariat of Science, Technology and the Environment - SECTMA (2009) - states that most of the water demand in the state of Pernambuco is for irrigation, with the Pontal river basin, GI-8 and Moxotó basin standing out, where demand for irrigation represents more than 90% of the basin's total demand.

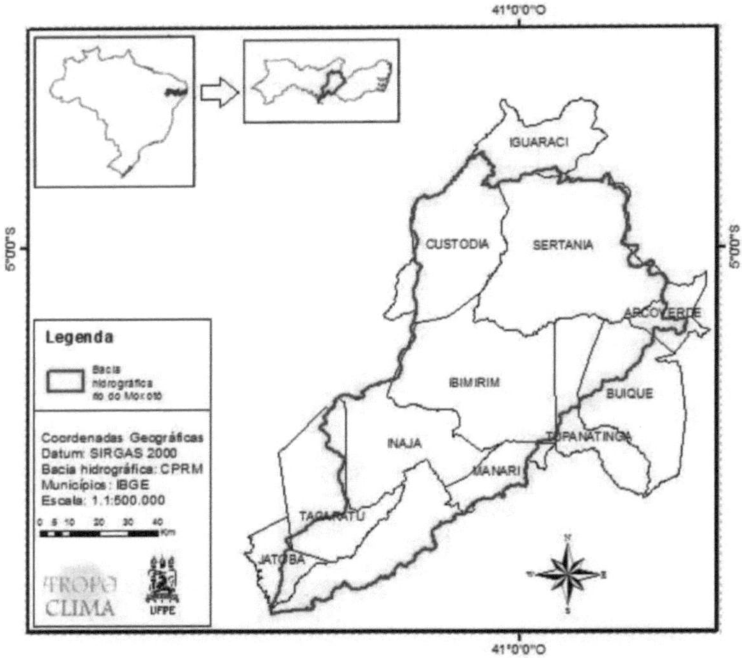

Figure 3: Location map of the Moxotó Hydrographic Basin - PE Source: Prepared by the author (2016)

The Moxotó river basin is covered from north to south, 89% of its area, by caatinga. It is considered a biodiversity *hot spot* due to the richness of its fauna and flora. The cyclical culture occupies 7% of the basin, while the remaining 4% is occupied by open tree vegetation (MELO, 2010).The geology of the Moxotó basin is complex. The headwaters are located on crystalline structures associated with the western slope of the Borborema plateau. At the beginning of the middle course of the main river, drainage headwaters appear on tabuliform and homoclinal morphologies associated with the plateau of the Jatobá basin. Following the same stretch, the river runs through a subsequent depression zone installed on the linear structures that separate the *Horst* from the east of the basin, from the plateau in a homoclinal structure that falls to the southeast of the western sector. In the lower section, the main river epigenetically cuts through a narrow *Horst* in an E - NE - WSW direction structured on sandstones of the Tacaratu Formation. The epigenyme identified seems to be an ancient capture of the middle course of the Moxotó River by the current course, whose most notable morphological feature is the narrow gorge superimposed on the *Horst* mentioned above (OLIVEIRA, 2009).In the final stretch, the river inflects to the southwest and forms an alluvial plane that blends in with the low pedimented surfaces, carved out of the crystalline structure. The alluvial plane reaches its maximum width near the mouth, denoting a substantial rise in the base level, which is even noticeable in the clogging pattern of the mouths of the smaller drainages adjacent to the main channel (OLIVEIRA, 2009).The basin's altitude varies between 200 and 1100m (Figure 4), which directly influences land use and land cover.

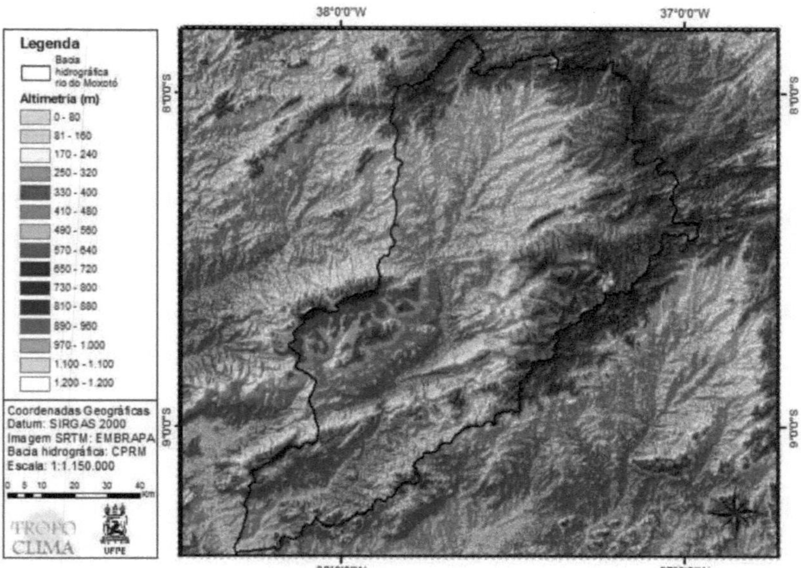

Figure 4: Altimetric map of the Moxotó river basin.
Source: prepared by the author. Climatic characterisation of the Moxotó Hydrographic Basin

It is known that rainfall in semi-arid areas varies from a minimum of 300 mm to 400 mm and a maximum of 700 mm to 800 mm (SEPLAN, 1993). According to Almeida (2001), rainfall is undoubtedly the element with the greatest monthly and annual variability in relation to most regions of the world.

The distribution of precipitation over the earth's surface is much more complex than the distribution of radiation or heat (air temperature). Part of the complexity in modelling the distribution of precipitation is attributed to the influence of certain factors, such as topography, distance from large bodies of water, prevailing air masses, among others (AYOADE, 1991).

In order to characterise the climate in the distribution of precipitation in the Moxotó basin, a bibliographic survey was carried out based on the main scientific publications on atmospheric systems, events and cyclical climate oscillations, anomalies and unidirectional changes that act in the region. Emphasising rain-producing systems and climate teleconnections such as those triggered by events like ENSO or the Atlantic Dipole.

Among the various climatic actors observed in the tropical region, the ones that stand out as contributing to the occurrence of rainy periods and droughts in the Brazilian semi-arid region are atmospheric systems: the ITCZ (Intertropical Convergence Zone), VCAN's (High-Level Cyclonic Vortices), DOL's (Eastern Wave Disturbances) or simply Eastern Waves and the Mesoscale Convective Complexes (MCC); the climatic events: El Nino - La Nina (ENOS) and the Atlantic Dipole.

The El Nino Southern Oscillation - ENOS is a cyclical climatic event that has effects on a global scale, with the most affected areas being South America, Indonesia and Australia. ENOS originates from the warming of the waters of the tropical Pacific Ocean, close to Australia and Indonesia, altering atmospheric pressure in different zones due to the change in direction of the speed of the trade winds and the displacement of rainfall from the tropical region to the coast of South America (NÓBREGA, 2000).

ENOS is one of the main climatic phenomena responsible for the long periods of drought in the semi-arid region of Brazil. By interfering with large-scale climatic characteristics in the Earth's atmosphere, it modifies rainfall patterns in this region. Its occurrence is due to the difference in Sea Level Pressure (SLP) between the central-eastern and western sectors of the Tropical Pacific Ocean, related to an anomalous warming of generally colder waters on the eastern side of this Ocean, which causes changes in the general circulation of the atmosphere, thus causing climatic alterations in various regions of the planet. (ALVES and REPELLI.1992).

According to Nóbrega (2000), the ENOS associated with the Tropical Atlantic Dipole directly affects the positioning of the ITCZ, thus influencing rainfall distribution over the Atlantic basin and northern South America.

The Atlantic Dipole presents Sea Surface Temperature (SST) anomalies with opposite signs to the north and south of the equator over the Tropical Atlantic, thus influencing the latitudinal position of the ITCZ, which in turn alters the seasonal distribution of precipitation over the Equatorial Atlantic, the Brazilian Northeast (NEB) and even the Amazon. It is considered that in the positive phase of the dipole there is below-normal rainfall throughout the NEB, while in its negative phase there is above-normal rainfall.

Thus, coincidentally, in the years when ENOS occurs with the rainy season over the NEB, and consequently in the semi-arid region, the atmospheric and oceanic phenomena over the Tropical Atlantic basin are more statistically significant for the inter-annual variability of the climate over this region than those of the Pacific itself (NÓBREGA, 2000).

Coelho *et al.* (2004) consider the ITCZ to be an atmospheric system with an important influence over the tropics, because the rainy season depends heavily on its position. In this sense, variations in the position of the ITCZ can result in periods of drought or intense rainfall over the semi-arid region, due to the occurrence of the positive phase of the dipole or the negative phase of the dipole, pushing the ITCZ further north or further south respectively.

Another important climate system that acts in the semi-arid region, especially during the austral summer months, are the so-called high-level cyclonic vortices (HLCVs). VCANs are synoptic-scale systems associated with heavy rainfall and their formation requires a frontal system in the Southern Hemisphere and another in the Northern Hemisphere close to equatorial latitudes and the confluence of the trade winds. The formation of VCANs is related to the Bolivian High and the Northeast Trough, as presented by Kousky and Gan in 1981. Below the 200 hPa trough, where the cyclonic vorticity intensifies, the Mid-Latitude Troughs (MLTs) and the Frontal Systems (FS) also act, causing both regular and irregular displacements in the direction of the vortex (KOUSKY and GAN, 1981). Since VCAN's are considered a closed cyclonic system, their action has the following characteristics: when their edges are over the semi-arid region, they cause rainfall, while when their (cold) centre is over the continent, they cause drought (veranicos). It can also be seen that this phenomenon occurs with a high frequency, since out of six years of observations, VCANs were present in five (SILVA ARAGÃO *et. al,* 1996).

Another climate system that acts on the semi-arid region, causing precipitation due to disturbances in the easterly winds, are known as easterly waves or simply EWDs (East Wave Disturbances). They form between the subtropical highs and the ITCZ (Intertropical Convergence Zone) and are dominated by strong and constant air currents close to the surface, called trade winds. In the lowest layer of the trade winds, the air is generally humid and the temperature is determined by heat exchange between the ocean-atmosphere interface. Above this layer is the Trade Winds Inversion Layer, where the temperature increases with height. Although the trade winds are related to stable weather conditions, in some regions of the tropics these situations are interrupted by disturbances which, because they move westwards and present wave-like configurations in the wind and pressure fields, are called easterly waves (NORQUIST et *al.,* 1977; OLIVEIRA, ALVES, and NOBREGA, 2010).

Since the 1980s, a large number of scientific articles have documented the environments, life cycles and observational characteristics of Mesoscale Convective Systems (MCS). These systems are important because they produce a large fraction of local-scale precipitation during the warm months and are often associated with severe weather. MCCs are clusters *of cumulus nimbus* covered by a dense layer of cirrus clouds that can be easily identified on satellite images due to their roughly circular shape and explosive growth of more than six hours (ALVES, 2011).

Knowledge of the characteristics of MCCs is of paramount importance since they contribute to the local and global hydrological balance. Zipser (ef *al.,* 1997, *apud* LAING E FRITSCH, 2006), using six years of data from the TRMM satellite, mapped the most extreme cases of severe convection on the globe, emphasising southern South America as the area with the most intense storms on Earth.

2.1.4. Causes of socio-environmental vulnerability in the Moxotó basin

The Moxotó river basin has always been marked by poor and precarious management. The basin has a history of management problems that have obviously directly affected the municipalities in the basin. The local economy is dependent on natural resources for the development of the main activities and not having efficient management has meant a state of economic recession for these municipalities, one of the factors responsible for their low municipal development indices (MELO, 2010; LIMA, 2011; LIMA E GATTO, 2013). However, this study focused on the case of the Moxotó Irrigation Perimeter because it has had a major social, economic and environmental impact on the basin.

Located in the Sertão physiographic zone of Pernambuco, part of the Sertão do Moxotó homogeneous micro-region, 334 km from Recife, the Moxotó Irrigated Perimeter (PIMOX) covers areas in the municipalities of Ibimirim and Inajá, with a total area of 12,395.96ha (6,375.00ha given over to irrigators), immediately downstream from the Eng°. Francisco Saboya (Poço da Cruz), extending for around 40 kilometres on both sides of the Moxotó River and comprising 565 agricultural plots (UNIVALE/DENOCS, 2013).

Although the basin is located in a region affected by climatic phenomena that cause long periods of drought, this is not the main factor causing vulnerability in its area. The main economic activity developed in the municipalities that make up the basin is agriculture, especially family farming. Although it has land with potential for irrigation - and even though it was the scene of development projects in the 1950s - some projects to develop the local economy were unsuccessful and caused impacts on the environment and the population, consequently contributing to vulnerability in the Basin.

In order to understand the socio-environmental vulnerability of the Basin, historical information was sought on the events that contributed to the dynamics of vulnerability in the area, namely: the construction of the Francisco Sabóia Public Dam or Poço da Cruz Dam and the construction of the Moxotó Irrigated Perimeter (PIMOX).

The Poço da Cruz weir was designed by the Federal Inspectorate of Works Against Droughts - IFOCS (now the National Department of Works Against Droughts - DNOCS). It is located in the municipality of Ibimirim - PE and has a maximum depth of between 37m and 50m. It is part of the Moxotó Hydrographic Basin and has direct contributions from three other rivers: Cupity, Mel and Piutá.

Construction of the reservoir began in 1957 and was quickly completed in 1958. Its main purposes were irrigation of the land downstream and upstream of the valley, fish farming and power generation. The latter would be achieved through the installation of a small hydroelectric power station. The plant that has been built has a total capacity of 1,660 kVA. However, the small hydroelectric power station has been deactivated for years (VASCONCELOS, 2009).

It was only at the beginning of the 70s that the water from the reservoir began to be used for irrigation, with its peak in the 80s. However, due to the lack of rainfall and irregular use of the resource, rationing measures had to be applied. The total collapse at the end of the 1990s led many farmers to abandon the irrigable areas and the entire structure of the irrigation perimeter (VASCONCELOS, 2009).

The Moxotó irrigated perimeter (PIMOX) was set up only in the municipalities of Ibimirim and Inajá. It is located downstream of the reservoir, and stretches for around 40 kilometres on both sides of the Moxotó River, covering a total area of 12,395.69 hectares, with an irrigable area of 8,596 hectares and an actual implemented area of 5,607 hectares. Diffuse irrigation covers 400ha of the area upstream of the reservoir and 600ha downstream. The aim of building the reservoir was to subsidise water for the irrigated perimeter and boost agricultural activity in the region. Most of the perimeter's structure is on alluvial soils, as they have a fine texture and are suitable for irrigation. The project was planned for 560 family plots with an average area of 6.6ha and 33 business plots with a total area of 3.03ha

(VASCONCELOS, 2009).

Most of the area was irrigated using the gravity method, which applies the technique of infiltration through furrows and siphons. The water from the weir is transported by a main canal that branches off into two separate adductors to supply the canals, however all the canals are open, which causes significant loss during transport due to evapotranspiration.

Between 1980 and 1983 PIMOX reached its peak, benefiting more than 6,000 families and generating more than 20,000 direct jobs and approximately 5,000 indirectly. Some of the state's newspapers even called the region the "golden hinterland of Pernambuco" (VASCONCELOS, 2009).

In 1991, a long period of drought began in the semi-arid region, leading to the rationing of the reservoir's water. In 1993, the volume of the reservoir reached critical levels. The total volume was reduced by 15% of its capacity, to around 74 million cubic metres, which led to the implementation of measures to restrict its use.

With the aim of controlling water wastage and improving the administration of the perimeter, in 1995 the General Assembly of Irrigators approved the creation of the Moxotó Valley Irrigating Rural Producers Association (UNIVALE).

The long dry season, high evapotranspiration rates and unsustainable use of the reservoir's water have considerably reduced its capacity to supply water to meet demand. This resulted in the total closure of the floodgates, which at the time of UNIVALE's creation were kept open constantly and without any control.

However, even with the application of water supply restriction measures and the reduction of irrigated areas, there was no resolution to the reservoir's volume problems or the impacts caused by it. This situation led to the total paralysis of the project and the suspension of the use of the reservoir's water for irrigation, aggravating socio-environmental and economic impacts.

The irrigation system was paralysed until 2004, when PIMOX resumed its activities, this time covering only 58 agricultural plots and 79 irrigators. The number of people benefiting from the system was reduced, since the system's structure didn't allow for the release of volumes greater than 1,000/s and the irrigation channels were still damaged, leading to water loss (VASCONCELOS, 2009).

If we add climatic factors to these unsustainable land use and water resource practices, these are the main factors that contribute directly to increasing the socio-environmental vulnerability of the basin.

2.2. CONCEPTUALISATION AND METHODOLOGY FOR ASSESSING

VULNERABILITY TO CLIMATE CHANGE: CREATION OF THE INDEX OF SOCIO-ENVIRONMENTAL VULNERABILITY TO CLIMATE CHANGE - IVMC

The process of generating information to build indicators from raw data also requires field knowledge. Indicators are used to build conceptual models of the functioning of a given system based on logical reasoning and/or the experience acquired between researchers and key players.

The conceptualisation of a socio-environmental system where one of the danger factors is possible climate change requires a quantitative model where phenomena can be converted into concrete values. It is therefore necessary to establish parameters to quantify the factors that generate vulnerability, such as the physical characteristics of the landscape and socio-economic characteristics. This information served as a reference for determining the levels of exposure, sensitivity and adaptive capacity of the system being analysed.

The raw data used to construct the indicators were selected according to the following criteria: they should be directly related to the hazard (CONNOR & HIROKI, 2005; MENDOZA, *et. al.,* 1999; MOSSET *et. al.,* 2000) and fit in with the scale of analysis and the specific characteristics of the Moxotó river basin (FOWLER *et. al.,* 2003; SULLIVAN & MEIGH, 2005).

The indicators were used to organise and systematise the information for planning, assembling and analysing the index that contributed to mitigation management. In addition to having physical, economic and social aspects, they should also have aspects that are directly related to climate characteristics and that are sensitive enough to reflect the changes in the system studied (Figure 5).

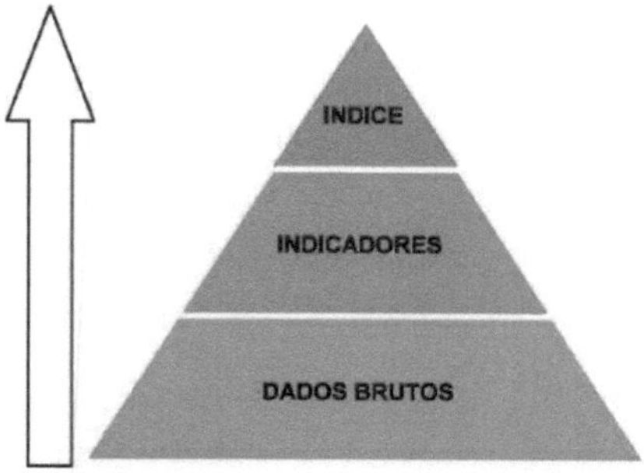

Figure 5: Hierarchisation of information for drawing up the index based on indicators and raw data.Source: adapted from CENDERO (1997).

In order to reduce the degree of uncertainty, only a few indicators were selected, but those that were selected present precisely the characteristics of the factors being analysed, namely exposure, sensitivity and adaptive capacity.

2.2.1. CONCEPTUAL MODEL

a. Construction of the Socio-environmental Vulnerability to Climate Change Index

At this stage, it was decided to work with information that was easily accessible and up-to-date, available to the general public and that was available on a municipal scale.

The main source of data extraction was the Brazilian Institute of Geography and Statistics - IBGE, which is the public body responsible for collecting and processing data related to the subject, as well as being the institute responsible for the national Census and the Pernambuco State Planning and Research Agency - CONDEPEXFIDEM. Data was also extracted from the Pernambuco Water and Climate Agency (APAC) and from the results of processing LANDSAT satellite images.

Raw data was collected from seven municipalities that form part of the Moxotó basin.

The indicators selected to make up the index were twelve, grouped into each corresponding sub-index: three to characterise the exposure sub-index, five to characterise the sensitivity sub-index, and four to characterise the adaptive capacity sub-index.

This index was constructed based on the methodology proposed by Monterroso (2012). The indicators that make up the sub-indexes were modified in order to better adapt the index to the reality of the Brazilian semi-arid region. Thus, the sub-indexes that make up the index were organised as follows:

i. EXHIBITION

This sub-index refers to the degree of climatic stress on a particular unit of analysis; it can be represented by changes in climatic conditions or rates of climatic variability, including the magnitude and frequency of extreme events. Some natural phenomena present themselves as threats and correspond to extreme phenomena such as floods, droughts, fires, storms, among others. Risk is established by the possibility of society being exposed to any extreme event (MARANDOLA *et. al.*, 2009). The following indicators are suggested to make up this sub-index:

> Rainfall trends: a threat to the primary activities prevalent in the semi-arid region is high climatic variability. This region, which has historically been hit by

extreme events, thus falls victim to frequent periods of drought. In regions with these climatic characteristics, agricultural production, livestock and hunting are more exposed to impacts that consequently also affect the social groups involved in such activities and the local economy itself. To build this indicator, monthly and annual rainfall data from 1930 to 2014 was analysed in order to identify trends in climate change, especially positive, negative or neutral changes in the basin's rainfall dynamics.

> Degraded soil rate: This sub-index considers the percentage of areas without vegetation cover. Vegetation provides protection against soil erosion, as well as controlling the albedo of solar radiation, which is higher in the absence of vegetation cover. The removal of vegetation cover in the semi-arid region, whether through burning or illegal logging, contributes to and potentiates soil exposure processes. The absence of vegetation cover

also contributes to soil leaching, i.e. the extraction or solubilisation of the chemical constituents of a rock, mineral, soil, sedimentary deposit, etc., by the action of a percolating fluid (DEMATTÊ *et al.,* 2004). To compose this indicator, it was decided to extract the information from the results of the Normalised Difference Vegetation Index (NDVI).

> Percentage of temperature change: Corresponds to the result of applying the B2 scenario (least critical) of the PRECIS model *(Providing Regional Climates for ImpactStudies).* The data for this indicator was taken from the IPCC's fifth report on climate change (IPCC, 2014).

ii. SENSITIVITY

Sensitivity is the state a system is in when it is potentially affected and modified by an internal or external disturbance, or a group of them. Measuring sensitivity determines the degree to which a system has been affected, making it possible to quantify a system's level of sensitivity. This sub-index is made up of indicators that make it possible to assess the social and environmental sensitivity of the Moxotó Basin system (MONTERROSO, *et. al.,* 2012).

Certain social groups are more exposed than others to environmental risks. The most sensitive social groups are poor communities in urban centres who suffer from the direct effects of air pollution and the impacts of the greenhouse effect; as well as family farmers in arid and semi-arid areas, who are subject to irregular rainfall patterns and frequent and long periods of drought. A wide variety of social and economic factors play a direct and indirect role in human sensitivity to climate change, including poverty, social inequality and lack of access to basic natural resources (CUTTER, 1996).

- Population - Incidence of poverty: Poverty here is conceptualised from the perspective of lack and/or absence of minimum resources. The conceptualisation of poverty is no longer limited to entry to a certain social level or income, but to the process by which people achieve (or not) their well-being (ESPÍNOLA, *et* al,.2010). The UNDP (2014), following Amartya Sen's proposal, also considers people's needs and/or deprivations due to their social condition as the final reference for determining poverty. In other words, in addition to family income, they consider access to education, health and quality of life indices (access to electricity, basic sanitation, drinking water, etc). Based on a multidimensional analysis criterion, it is understood that poverty has a relative and extremely complex concept that is based not only on the lack of financial resources, but also on the lack of real opportunities for development, determined by the limitations imposed by society (UNDP, 1997). Poverty is therefore recognised as an extremely important indicator in measuring vulnerability to climate threats. Low-income social groups have a lower capacity for control than other groups and therefore suffer disproportionate consequences (MARANDOLA et. *al.,* 2005).
- Health - Number of public health facilities: This topic includes the lack of access to public health treatment as a sensitivity indicator. Since a given area suffers from various impacts, these directly affect people's lives and sometimes cause similar illnesses and impacts. It is therefore understood that the smaller the number of public health facilities, the more difficult it is for low-income people to access health treatment, thereby increasing their vulnerability (UNDP, 2014).
- Sanitary conditions - Number of homes without adequate basic sanitation: The World Health Organisation (WHO) defines basic sanitation as "the control of all factors in man's physical environment that have or can have a deleterious effect on physical, mental and social well-being". The aim of basic sanitation is to ensure human health, bearing in mind that many diseases can develop as a result of poor sanitation (UNDP, 2014).
- Economic activity - Number of workers in primary activities (agriculture, livestock and hunting): Primary activities are directly dependent on natural factors and once they undergo changes and impacts, these activities would be the first to undergo

transformations. Primary activities are predominant in the municipalities that make up the Moxotó basin and if there are changes in the natural system, there will be changes in the basin's socio-economic system. According to Tavares (2010), changes in geosystems can lead to changes in socio-economic systems and vice versa. As the formatting of urban and rural spaces, linked to financial interests, has prevailed and produced rapid changes in the flows of matter and energy within the socio-economic systems, with repercussions on the geosystems, there is speculation about a tendency towards climate change of anthropogenic origin that would produce impactful consequences that feed back into both systems.

iii. ADAPTIVE CAPACITY

A system's ability to cope with the consequences of climate change impacts contributes to the process of mitigating vulnerability, making it necessary to implement measures that help minimise impacts and techniques that enhance the adaptive capacity of social groups. The adaptive capacity of a social group reflects its flexibility in modifying characteristics and behaviours to better cope with difficulties, with the aim of monitoring the factors that drive change (MONTERROSO, et. al., 2012). The following indicators were selected to make up this sub-index:

> Human capital - Literacy rate: Literacy is understood as "the ability to use graphic symbols that represent spoken language so that the collective knowledge of the group is externalised and perpetuated in time and space" (BATISTA, 2006, p. 15). Literacy is mainly related to the human capacity to use a set of techniques to decode and produce written material, which, according to McGarry (1999), refers to print literacy or basic literacy. For Bawden (2002), the concept of literacy goes far beyond mastering reading and writing. For him, understanding and interpreting space are fundamental and indispensable in the literacy process. Still according to Bawden, in recent decades the concept of literacy has been complemented by so-called "skills literacy", this broader concept of literacy having been developed to refer to the technological and complex information that is at an all-time high. Bawden concludes by saying that a variety of names could be given to literacy, depending on the skill to which they refer: agricultural literacy, film literacy, digital literacy, among others. Literacy is an important indicator that contributes to adaptation, since people who better interpret their surroundings can better predict situations, such as the work of

the 'prophets of the sertão', farmers from the Sertão do Cariri who meet annually to share weather forecasts for the current year. Thanks to their mastery of reading, writing and interpretation, these farmers can prevent the effects of climatic phenomena (ALVES, 2006).

> Household economic capital - Number of permanent private households with nominal monthly household income classes of more than 2 to 5 minimum wages: Undoubtedly all humans have some level of vulnerability to environmental impacts. People's ability to adapt to and control change varies. However, social groups with lower financial incomes have a lower capacity to adapt to these changes and are more vulnerable to the negative effects of climate phenomena, as well as other threats (IPCC, 2001).

> Public financial capital - public investment in social programmes and infrastructures: this indicator refers to the role of the state as the main actor in preventing impacts by planning and implementing mitigation actions that foster ways for the population to better adapt to the adversities of climate change. Therefore, it is understood that investing in social programmes that directly contribute to the population's well-being improves their adaptive capacity (UNDP, 2014).

> Natural capital - Availability of water per household: Access to water is a problem in every arid and semi-arid region in the world. Water is the fundamental natural resource for human survival. Having access to this resource means being better able to cope with the misfortunes that climate change can cause. Therefore, in this topic, the number of households with access to piped water will be considered as a contribution (UNDP, 2014).

iv. DATA PROCESSING

To draw up the index, a database was created with all the information on each of the indicators, in which the maximum value of each indicator would reflect its maximum severity, with the exception of the indicators that make up the adaptive capacity sub-index, in which the maximum value presented by the indicators would reflect their greatest positive potential.

In order to be able to compare all the resulting values with each other, all the indicators were standardised to eliminate scale differences. To do this, the following equation (1) was applied:

$$Z_v = (X_i - X) / DS$$

(1)

Where Z represents the standardised value of the variable v; Xi is the observed value of v, and X is the average value of the set of values i, and SD is the standard deviation of the set of values i. After standardising the indicators, we proceed to calculate each of the sub-indices: exposure, sensitivity and adaptive capacity.

The standardised values of the indicators for each sub-index are algebraically added up using the following equation (2). Once the sub-indices have been calculated, the scale of values from 0 to 100 is normalised.

$$L_p = (I_{(obs)} - \text{minimum } I) / (\text{maximum } I - \text{minimum } I)$$

(2)

Where I is the sub-index of variable p, Iobs is the sum of the indicators of each sub-index. Minimum I is the minimum observed value of the set of Ip values; and maximum I is the maximum observed value of the same set of Ip values. After this stage, all the resulting values were transformed into percentages.

With all the values calculated and duly standardised on a percentage scale. The index is finalised by calculating the vulnerability of each municipality in the basin using equation (3).

$$\text{Vulnerability} = (\text{exposure} + \text{sensitivity}) - \text{adaptive capacity}$$

(3)

Where vulnerability is equal to the sum of the standardised exposure and sensitivity results and subtracted by the adaptive capacity values.

The municipal scale was chosen for spatialisation and construction of the vulnerability index. The criterion for selecting the type of scale was according to the availability of information on all the indicators and all the municipalities that make up the basin. The state of Pernambuco, through CONDEPE - FIDEM, updated the 2010 municipal census database, making it the most up-to-date data source available for this scale.

CHAPTER 3

RESULTS

3.1. SOCIO-ENVIRONMENTAL VULNERABILITY IN THE MUNICIPALITIES OF THE MOXOTÓ RIVER BASIN

Vulnerability is, in a very simple and summarised definition, the potential for loss of a given system. Basically, vulnerability is made up of three factors: exposure to risk, which is determined by the circumstances that place social groups or physical locations in front of an imminent risk; sensitivity, which translates into the degree to which a system can be affected, as well as its capacity to respond (positively or negatively) to danger; and adaptive capacity, which is the set of actions that increase or reduce the capacity of social groups and physical systems to face and recover from environmental threats. These simple and generic definitions are the essence of vulnerability (CUTTER, 1996; WISNER *et al.*, 2004; ADGER, 2006, BIRKMANN, 2006).

Identifying the vulnerability of a particular social group or system requires a systemic approach, capable of explaining the complex interactions between social, natural and artificial systems, as well as the importance of understanding the dynamics of each of these systems and the natural processes that give rise to risks. In this way, it is a question of getting as close as possible to understanding the impacts of these natural processes or events.

Environmental disasters occur when natural threats affect the most vulnerable social groups. In this context, Romero and Mendonça (2012) highlight the "pressure and release" model proposed by Blaikei et *al.* (1996). According to the author, vulnerability is directly linked to social processes and disasters, in turn, are found at the intersection of opposing forces: social processes that generate vulnerability and physical exposure to a natural threat. The 'pressure' corresponds to the tension exerted by both forces on the population and the 'release'

refers to actions that have the direct and indirect aim of reducing vulnerability.

Research into vulnerability helps to identify how a given event, or set of events, can produce different impacts in the same affected region. In addition, identifying vulnerability provides tools for investigating social factors that interfere with nature, redistributing the pre-existing risk of an event and the damage afterwards. However, comparing the differences in vulnerability generated in the same region, or between regions, by the same event, or a set of them, requires a consistent concept of scale, which is fundamental when discussing the results. For analysing vulnerability, systematic comparison between areas is a fundamental component of the study (CUTTER, 2011).

Studies that aim to identify and analyse socio-environmental vulnerability must prioritise and highlight the relationship between environmental changes and permanent vulnerability, i.e. avoid considering that environmental changes, extreme events and/or natural disasters are solely responsible for the extraordinary disturbances that interrupt the normal behaviour of natural processes, and cannot analyse them separately from the social, economic and cultural characteristics of the social group affected. According to Romero and Mendonça (2012), in vulnerability studies, analysing disasters separately from social elements is the same as systematically ignoring the risks of everyday life, especially in the face of natural threats. Poverty and lack of access to health and education are factors that increase the capacity to absorb impacts. According to the authors, disasters have multidimensional scales and occur between social and natural systems.

Social vulnerability therefore represents the inclination of social groups to the negative impacts of hazards and disasters (CUTTER et *al.,* 2003). It is vulnerability that is responsible for identifying the factors in society that increase or decrease the capacity to prevent or promote actions to recover from the impacts of disasters. It is also the study of vulnerability, with the help of cartography, that helps to understand the distribution of risks and potential damage, i.e. it highlights the relationship that exists between vulnerable societies and vulnerable natural environments (CUTTER, 2011).The atlas of social vulnerability of Brazilian municipalities (2015), produced by IPEA, states that in 2010, 47.7 per cent of municipalities in the Northeast were in the high range and 32.4 per cent of municipalities were in the very high range of the social vulnerability index. However, comparing the results of the social vulnerability index, both in 2000 and 2010, Pernambuco was the second to show better results in reducing social vulnerability. However, it still has many municipalities with very high social vulnerability, especially in the Moxotó micro-region,

Some of the municipalities that make up the Moxotó basin were visited on a fieldwork basis. The visits took place in 2012 and 2013. The municipalities that were part of the *in situ* activity itinerary were: Arcoverde, Buíque, Ibimirim, Inajá, Tacaratu and Tupanatinga. Brazil's semi-arid region is deeply affected by climate variability, above all due to the effects of *El Niño* and the Atlantic Dipole, which in this region translate into long and severe periods of drought. During the fieldwork, it was possible to see the impacts caused by climate adversity. In addition to the impacts of drought, it was also possible to identify predatory activities such as burning and the extraction of vegetation. The methodological procedure for the fieldwork was basic and restricted only to identifying the area, taking photographs of daily life in the municipalities visited and of predatory activities such as burning and vegetation extraction. Photographic records of the landscape were also taken to characterise how the landscape behaves during the dry season. In the first fieldwork, in 2012, the municipalities of Arcoverde, Buíque and Ibimirim were visited. These municipalities are located in the eastern sector of the basin, with the exception of the municipality of Ibimirim, which expands its territory to the extreme western edge of the basin, also covering the entire central region. The eastern edge of the basin is where, according to the analysis of vegetation cover, the permanent native vegetation is concentrated, with vegetation at the highest altitudes.During 2012, these municipalities, as well as the entire semi-arid region, were impacted by drought, marking the beginning of a long period that lasted until the end of 2015, as a result of the long dry season. The images shown below were taken in May and June 2012. The first images correspond to the native caatinga in the municipality of Arcoverde, which decreases considerably in density during periods of drought, favouring exposure of the soil (Figure 6 and 7).

Figure 6: native vegetation in the municipality of Arcoverde (November 2012).
Photo: Keyla Alencar (2012).

Figure 7: Native vegetation in the municipality of Arcoverde. Photo: Keyla Alencar (2012).

According to Christofoletti (1999), from a physical perspective, resilience is the ability of a body to recover its original shape and size after being subjected to a stress that does not exceed the limit of its elasticity. Adapted to environmental systems, resilience determines the persistence of the system's internal dynamics, reflecting its capacity to absorb changes and, consequently, producing readjustments in the final state when compared to the initial conditions. In other words, the system maintains its structure and characteristics, although it doesn't always maintain the same values in the attributes that make up the system. Also according to the author, areas subject to strong climatic fluctuations, such as the Brazilian Northeast, and obviously the semi-arid region, suffer significant decreases in fauna and flora during the dry phases, although they recover quickly after the disturbances.

In the municipality of Buíque, in addition to the dry and sparse vegetation, predatory activities involving the extraction of native vegetation and forest fires were also recorded (Figures 8, 9 and 10). Artisanal constructions to dam water were also found along the way. Some households receive help from social programmes that install cisterns (water reservoirs). The ASA - Articulation of the Brazilian Semi-Arid - together with the Government of

The state of Pernambuco is responsible for distributing and installing the cisterns, but they are insufficient to meet all the demand in the semi-arid region. Farmers who don't have cisterns or artesian wells look for alternatives to conserve water.

Figure 8: vegetation removal in the municipality of Buíque.
Photo: Keyla Alencar (2012).

Figure 9: Removal of vegetation by burning in the municipality of Buíque.
Photo: Keyla Alencar (2012).

Figure 10: Outbreak of forest fire in the municipality of Buíque.
Photo: Keyla Alencar (2012).

The problem with artisanal dams is that they are inefficient and don't conserve water for many days, especially in an area where evaporation rates are high. According to reports from the owners of the small dams, the water that remained stagnant lasted a maximum of two to three weeks and was of no use for human consumption, being used only to feed animals (Figures 11 and 12).

Figure 11: artisanal construction for damming water in the municipality of Buíque.
Photo: Keyla Alencar (2012).

Figure 12: artisanal construction for damming water in the municipality of Buíque. Photo: Keyla Alencar (2012).

Despite incentives from the Pernambuco state government to provide drinking water to all households in the Semi-arid region, according to FIDEM (2013) none of the municipalities that make up the basin have more than 50 per cent of households with piped water. This reality is reflected in the low capacity of the population of these municipalities to cope with the periods of drought that are so frequent in the region, which makes them look for alternatives, such as artisanal dams, which are not always effective in mitigating the problems generated by drought.

The third municipality visited was Ibimirim, also badly affected by the drought. The area of native vegetation was reduced and the Puiu lagoon, the municipality's second main water reservoir after the Poço da Cruz reservoir, dried up completely. During the fieldwork, the tragedy of countless animals dying from lack of water and food was recorded.
The Puiu lagoon is more than a kilometre long and more than 400 metres wide. It is surrounded by native vegetation such as caatinga shrubs as well as fruit trees, palm trees and subsistence crops.

Historically, the Puiu lagoon has always been the main water resource for the region's natives to cope with drought (Figures 13 and 14).

Figure 13: Panoramic view (low) of the Puiu lagoon. Photo: Keyla Alencar (2012).

Figure 14: Panoramic view (top) of the Puiu lagoon. Photo: Keyla Alencar (2012).

Ibimirim's main economic activity is agriculture and livestock farming. Both activities are directly dependent on climatology and suffer impacts on the local economy, which translates into losses in agricultural production, malnutrition and the death of animals (Figure 15 and 16).

Figure 15: Dead cattle, municipality of Ibimirim. Photo: Keyla Alencar (2012).

Figure 16: Malnourished cattle, municipality of Ibimirim. Photo: Keyla Alencar (2012).

The second fieldwork took place in 2013 and the municipalities of Inajá, Tacaratu and Tupanatinga were visited. The first municipality visited was Tupanatinga. All the watercourses in this municipality, like most municipalities in the semi-arid region, have an intermittent flow regime and a dendritic drainage pattern. Tupanatinga, like the other municipalities visited, had dry bodies of water. The Mel stream, one of the main tributaries of the Moxotó river, had low water availability (Figure 17).

Intermittent rivers and streams are present during the rainy season, but during the dry season they disappear. Intermittent rivers, also known as temporary rivers, are fed by surface and subsurface runoff. They temporarily disappear in the dry season because the water table becomes lower than the level of the channel, ceasing to feed them. Although this is a natural characteristic of the rivers and streams of the semi-arid region, it is this water condition and the lack of alternatives to accessing water that increase the population's sensitivity to facing the adversities of drought.

Figure 17: Stretch of the Mel stream, municipality of Tupanatinga. Photo: Keyla Alencar (2013).

Inajá was the second municipality visited, also heavily impacted by the lack of rain. During the visits to farmers, obvious material losses were recorded (Figure 18), as well as the death of numerous animals. The complete lack of water for irrigating crops was one of the population's main problems.

The damage that recent droughts have caused to animal production is directly related to livestock losses. According to the IBGE, 1.3 million cattle, 696,000 goats, 784,000 sheep and 755,000 poultry were lost in the Northeast in 2012, with losses also occurring in pig and horse herds. The most pronounced losses were in the northeastern states of Bahia (40 per cent), Paraíba (28 per cent) and Pernambuco (24 per cent).

According to the Special Report on the Water Crisis (2014), produced by the National Water Agency, by the year 2025 more than 70 per cent of cities with a population of more than 5,000 inhabitants in the semi-arid region will face a crisis in the supply of water for human consumption. In short, among the predicted impacts of climate change in the Brazilian semi-arid region are: changes in the vegetation of the Caatinga; a decrease in water in lakes, dams and reservoirs; greater vulnerability to torrential rains concentrated in a short space of time, resulting in floods and serious socio-environmental impacts; a greater frequency of consecutive dry days and heat waves; the unfeasibility of subsistence agricultural production in large areas and an increase in migration.

Figure 18: Dead cattle, municipality of Inajá. Photo: Keyla Alencar (2013).

The last municipality visited was Tacaratu. In this municipality, as well as in all the other municipalities visited, the adversities to which the population is exposed during periods of drought can be seen. Especially the rural population, which is mostly made up of family farmers who don't have the technology to plant crops and in many cases don't even have access to piped water (Figures 19 and 20).

Mendonça (2005) emphasises that climate change is aggravating the increase in common diseases in the tropics, such as malaria, dengue fever and leptospirosis. However, the author also draws attention to the lack of socio-economic planning, poverty and human misery (which, despite the many advances in social projects in the semi-arid region, still has high poverty rates), the intense process of urbanisation in cities and the formation of highly degraded socio-environmental conditions, all of which increase the health problems typical of the tropical zone.

Figure 19: Women washing clothes in a "bica", Tacaratu municipality.
Photo: Keyla Alencar (2013).

Figure 20: Women washing clothes in a "bica", Tacaratu municipality.
Photo: Keyla Alencar (2013).

The following topics correspond to the results of the exposure, sensitivity and adaptive capacity sub-indices, respectively, and finally the result of the climate change vulnerability index.

3.2. STATISTICAL AND SPATIAL RESULTS OF THE SUB-INDICES AND INDEX

3.2.1 Exposure sub-index

The internal dimension of vulnerability is made up of the system's sensitivity and adaptive capacity. The elements that make up the internal dimension relate to defence and security capacity, as well as the ability to anticipate, cope with, resist and recover from a given impact. The internal processes of the system include land use, social, economic, cultural, political and environmental factors. These processes have a decisive influence on vulnerability (SMIT, 2006).

However, this dimension is made up of exposure and risk to a given phenomenon or stressor. O'Brien *et al.* (2004) define exposure as the nature or degree to which a system is exposed to considerable climatic variations. Exposure thus relates to a system's level of climatic stress.

Exposure can be represented by changes in long-term climate variables, as well as changes in climate variability, including the magnitude and frequency of extreme events. In this study, to represent exposure, we considered changes in local climate variables by applying the method of analysing trends in precipitation, the percentage increase in temperature and the variation in vegetation density.

The result of the exposure sub-index is expressed as the arithmetic mean of each of the indicators shown below:

Table 9: indicator results for each corresponding municipality.

MUNICIPALITIES	RAINFALL TRENDS	DEGRADED SOIL RATE	TEMPERATURE VARIATION
Arcoverde	12,50%	58,16%	4%
Buíque	0,00%	67,66%	4%
Custody	-12,50%	43,88%	4%
Ibimirim	12,50%	58,89%	4%
Inajá	12,50%	32,95%	4%
Iguaraci	0,00%	18,54%	4%
Jatoba	0,00%	62,42%	4%
Manari	0,00%	56,15%	4%
Sertânia	0,00%	58,35%	4%
Tupanatinga	0,00%	72,46%	4%
Tacaratu	0,00%	39,99%	4%

Source: prepared by the author.

The rainfall trend indicator is represented by the result of applying the Mann-Kendall method to the 1930 - 2014 rainfall series (chapter 3). The rainfall data was obtained from the Pernambuco Water and Climate Agency (APAC). The result of this indicator corresponds to the average trend calculated by the Mann-Kendall index.

The degraded soil rate indicator is represented by the NDVI result (chapter 4) applied to satellite images from 1984 - 2011. And the result of this indicator corresponds to the percentage of areas with soil with any type of degradation corresponding only to the period analysed. The result of the temperature change indicator corresponds to the temperature increase value presented by the PRECIS model in the B2 scenario for the semi-arid region (Graph 23).

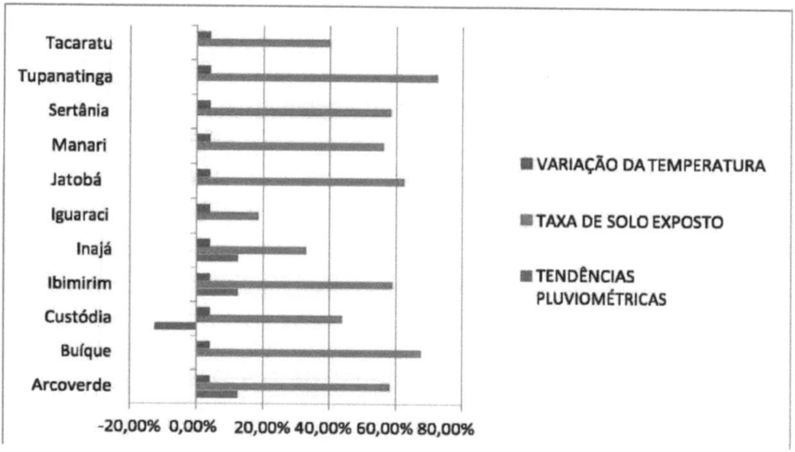

Graph 23: Graph of the results of the exposure sub-index indicators.
Source: prepared by the author.

It is important to emphasise that the indicators that make up the exposure sub-index represent three axes: changes in the basin's precipitation dynamics, the environmental problem represented by the absence of vegetation and future climate change represented by the future temperature change scenario. Each of the indicators presented shows the severity of the exposure. To reach the final result of this sub-index, the equation **Ip=Iobs-minimoI/maximoI-minimoI** was applied to the results of the indicators and the result was as follows:

Table 10: Results of the exposure sub-index

MUNICIPAL	EXPOSURE SUB-INDEX
Arcoverde	64%
Buíque	62%
Custody	22%
Ibimirim	65%
Inajá	33%
Iguaraci	-3%
Jatoba	56%
Manari	49%
Sertânia	51%
Tupanatinga	67%
Tacaratu	30%

Source: prepared by the author.

Once the exposure sub-index had been constructed with values assigned to each municipality in the basin, the results were prioritised by degree of exposure. The final result was divided into three categories according to frequency:

Table 11: criteria for the severity of climatic exposure in the basin's municipalities.

Exposure sub-index	0-40	41 -60	61 -100
Severity of exposure	Low	Average	High

Source: prepared by the author.

The final results were spatialised for better analysis (Figure 21). Thus, we observed that four of the municipalities in the basin had high exposure (61 - 100) to phenomena and stressors (Arcoverde, Buíque, Ibimirim and Tacaratu); another four municipalities had medium exposure (41 - 60) (Jatobá, Manari and Sertania); and four municipalities had low exposure (0 - 40) (Custódia, Inajá, Iguaraci and Tupanatinga).

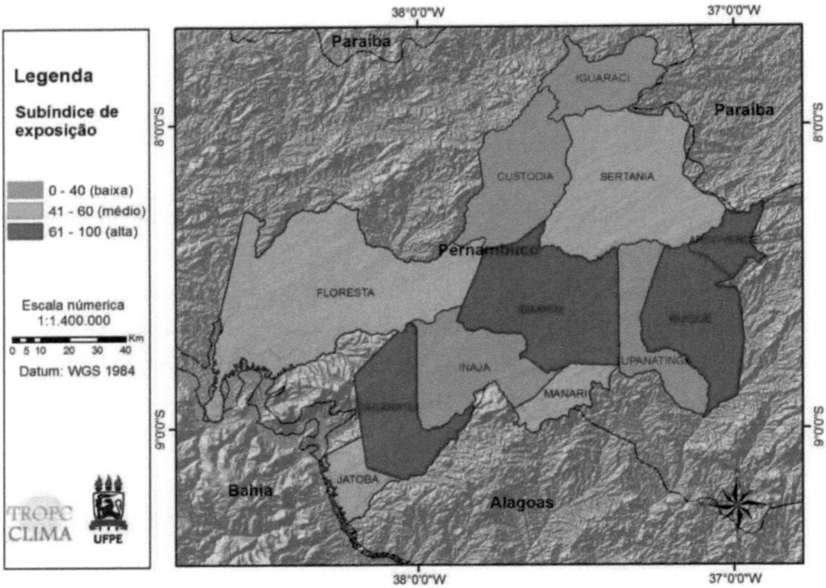

Figure 21: Map of the exposure sub-index.

Source: prepared by the author (2015).

According to the results of the exposure sub-index, it was concluded that: 25% of the municipalities that make up the basin suffer an increase in precipitation, 8.3% suffer a decrease in precipitation and the remaining municipalities did not show any type of trend. Only 33.3% of the municipalities have less than 50% exposed soil, all the others have more than 50% of their total area with exposed soil and, with regard to the increase in temperature, the value applied was the same for the entire basin, since it corresponds to the value given by the PRECIS model for the entire semi-arid region. This percentage represents an increase of 3° to 4° in the average temperature of the semi-arid region.

Exposure to threats can be minimised by reducing the risk, i.e. reducing the probability of exposure to the impact and, consequently, reducing the possibility of damage occurring. Theoretically, exposure can always be avoided by isolating people from dangerous situations, but this is an alternative that doesn't work in practice. Therefore, the most effective alternative for reducing exposure is to expand research into exposure factors, in this case climate change, and apply the knowledge from this research to environmental planning.

3.2.2 Sensitivity sub-index

The degree to which a system responds to a given change in controlling factors is represented by its sensitivity. The degree of susceptibility of a system is represented by sensitivity when conceived as the ability to react or the willingness to resent the slightest external actions, variations or disturbances (CHRISTOFOLETTI, 1999). Sensitivity also refers to the ability of an environmental system to cope with a disturbance, in this case climatic.

The concept of sensitivity is made up of three other concepts: stability, resilience and flexibility. Stability refers to a system's ability to remain stable during a disturbance; resilience represents the system's ability to recover; and flexibility represents the system's internal malleability. Sensitivity represents the aspects that constitute the resistance capacity of a given system and not a specific limit for collapse, not least because it is quite complex to define the limit of any system, whether natural or human (SMITHERS and SMIT, 2009).

This sub-index is made up of indicators representing economic activities, health and the profile of the population occupying the basin. The indicators are: number of workers in primary activities, number of private households without adequate basic sanitation, number of public health facilities and incidence of poverty.

The main economic activities carried out in the municipalities that make up the basin are directly or indirectly linked to the land. Agriculture, livestock farming and fishing (the latter mainly in the Puiu lagoon - Ibimirim) are activities that will be strongly affected by climate change, as will related activities, such as the 'fairground' workers who sell local agricultural products, which are common in the semi-arid region. The profile of rural workers in these municipalities is structured on the family level and in small cooperatives. There are no modern structures and few public and private initiatives to promote effective and efficient planning of these activities, which in times of drought, for example, leads to maximum losses and a strong economic impact on the municipalities.

According to the IBGE, during the drought of 2012, the Semi-Arid region lost 696,000 cattle and 784,000 goats and, in the poultry sector, 797,000 roosters, hens and chickens, as well as other animals. In total, the IBGE reports a loss of 137 million dead animals. Articulação do Semiárido (2013) also states that the negative impacts of the 2012-2013 drought have caused great damage, but that it is the sensitivity of economic activity and the lack of planning and public policies that have increased the impacts. The lack of political dimension and infrastructure programmes expose primary activities to impacts and thus to damage. The main people affected by the impacts were family farmers and smallholders, who are the majority in the region.

The indicators "number of private households without adequate basic sanitation" and "number of public health facilities" represent how prepared municipalities are to face a possible increase in diseases caused by climate change. The lack of basic sanitation is one of the main problems in the proliferation of infectious diseases. Mendonça (2009) states that, despite the obvious climatic influence on the emergence of epidemics of mosquito-borne diseases, the return of these diseases is a consequence of ineffective public health policies.

Brazil has already made considerable progress in reducing misery and poverty, especially in the semi-arid region. However, Obermaier and Rosa (2013) point to the need to interpret adaptation as a process and vulnerability as a dynamic condition, as well as the importance of considering poverty as a fundamental indicator when analysing vulnerability to climate change, since the poorest communities are the most exposed to the impacts of climate change and variability.

The data in this sub-index comes from CONDEPE/FIDEM - the Pernambuco State Planning and Research Agency. They are projections for 2013 based on the 2000 and 2010 censuses. All the social and economic indicator data used in the preparation of this sub-index - and in all the others - was extracted from this same database.

The sensitivity sub-index represents three axes: the population's purchasing and income profile; access to health services; and agriculture, livestock and hunting. The following projected results are presented:

Table 12: Indicator results for each corresponding municipality.

MUNICIPALITIES	INCIDENCE OF POVERTY	NUMBER OF PUBLIC HEALTH	NUMBER OF PRIVATE HOUSEHOLDS WITHOUT ADEQUATE	NUMBER OF WORKERS IN PRIMARY ACTIVITIES

			BASIC SANITATION	
Arcoverde	38,49%	0,67%	16,01%	10,36%
Buíque	38,93%	0,67%	15,94%	30,43%
Custody	32,41%	0,67%	34,05%	27,28%
Ibimirim	41,27%	0,33%	23,08%	36,93%
Inajá	41,15%	0,33%	26,03%	24,81%
Iguaraci	37,41%	0,33%	30,62%	29,63%
Jatoba	40,48%	0,33%	17,36%	36,99%
Manari	40,66%	0,33%	20,09%	35,65%
Sertânia	26,35%	0,33%	37,01%	31,83%
Tupanatinga	41,43%	0,33%	21,11%	33,56%
Tacaratu	42,68%	0,33%	21,04%	37,63%

Source: prepared by the author.

As mentioned above, the indicators are projected data for 2013. We preferred to use projected data because it is more up-to-date than census data. The indicators for the incidence of poverty, the number of private households without adequate basic sanitation and the number of workers in primary activities are projected according to the results of the 2000 and 2010 censuses. However, the indicator for the number of public health establishments refers to the total number of public hospitals per municipality and the percentage is given based on the total number of hospitals in the state of Pernambuco. As such, the municipalities with a value of 0.67% in this indicator have two public health establishments. The municipalities that scored 0.33% have only one public health establishment. For better interpretation, the results of the indicators were applied in graph form (Graph 24).

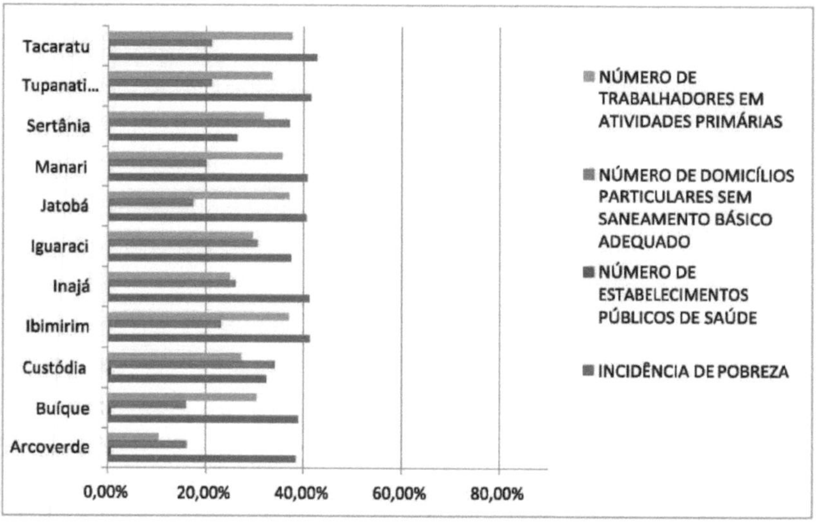

Graph 24: Graph of the results of the sensitivity sub-index indicators.

Source: prepared by the author.

The results of the indicators form the equation: **Ip=Iobs- minimoI/maximoI-minimoI** (the definition of the equation can be found in chapter 2), obtaining the result of the sensitivity sub-index, which can be seen below:

Table 13: Results of the sensitivity sub-index

MUNICIPALITIES	SENSITIVITY SUB-INDEX
Arcoverde	63%
Buíque	84%
Custody	92%
Ibimirim	100%
Inajá	91%
Iguaraci	97%
Jatoba	94%
Manari	96%
Sertânia	94%
Tupanatinga	95%
Tacaratu	100%

Source: prepared by the author.

As with the exposure sub-index, the result of the sensitivity sub-index was prioritised according to frequency, as can be seen below:

Table 14: criteria for the severity of climate sensitivity in the basin's municipalities.

Sensitivity sub-index	0-40	41 -60	61 -100
Severity of sensitivity	Low	Average	High

Source: prepared by the author.

None of the municipalities analysed showed a low result in this sub-index. All the other municipalities showed a high severity of sensitivity (Figure 22).

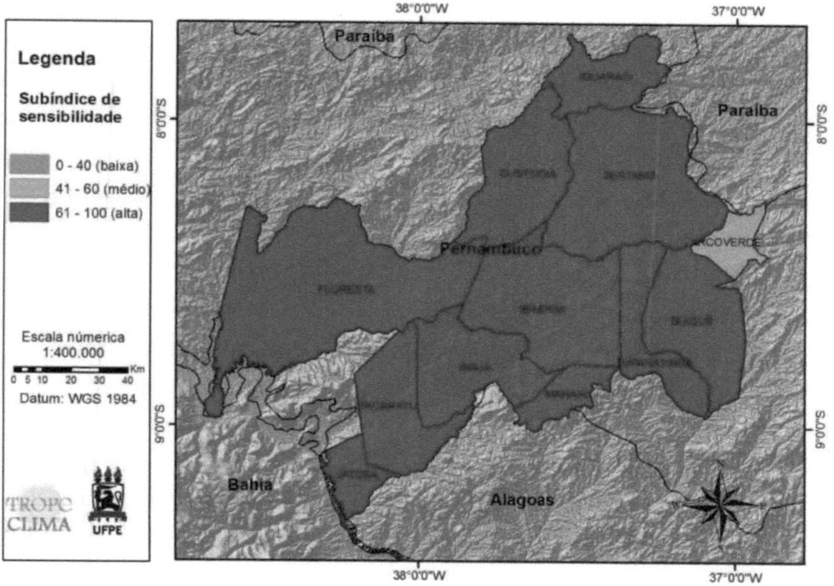

Figure 22: Map of the sensitivity sub-index. Source: prepared by the author.

Sensitivity to climate change was understood as the degree to which the municipalities in the basin, their population and their primary economic activities could be affected by changes and variations in the climate. It was observed that of all the municipalities analysed, only the municipality of Sertânia had less than 30% incidence of poverty; all the other municipalities have more than 30% of their population in a state of poverty.

Regarding the indicator of adequate basic sanitation, which is directly related to sanitation and health issues, the municipalities nevertheless have a deficit. Only 25 per cent of municipalities have more than 20 per cent of households with adequate basic sanitation (according to FIDEM's definition, adequate basic sanitation refers to households with a general sewage system or septic tank and rubbish collection).

According to the results of the indicator of workers in primary activities, only the municipality of Arcoverde had less than 15% of its population in this type of activity. It is important to clarify that this indicator represents not only the number of workers in this activity, but also the profile of primary activities in the basin, which are mostly small farmers and family farmers who do not have the infrastructure to produce during periods of drought, nor do they have the technical support to boost production. As for climate-dependent primary activities, any change or variation in the climate would directly affect these activities and, consequently, the local economy (BUAINAIN and GARCIA, 2013).

According to the results, the existing infrastructure for health and economic activities in the municipalities is insufficient to help reduce sensitivity. It would be appropriate to promote improvements in this area, especially in the implementation of thematic plans that accompany the dynamism of the local population. It is suggested that specific plans be implemented to deal with climate change, but above all to meet the needs of each municipality, taking into account the particularities of each population.

3.2.3. Adaptive capacity sub-index

Adaptive capacity forms part of vulnerability and is one of the factors that, together with exposure and sensitivity, characterise vulnerability. Adaptive capacity describes the ability of a system to adjust to expected conditions, i.e. the ability to cope with the adverse consequences of an impact, it is the ability of a system to modify its internal structure in favour of a less vulnerable condition (LUERS *et al.,* 2003). Adaptive capacity can be determined according to health, technology, education, information, skills, infrastructure, access to resources and management capacity (BROOKS*,* et *al.,* 2005; MCCARTHY, et *al.*, 2001; O'BRIEN et *al.,* 2004).

Two different approaches have been proposed to understand adaptive capacity. Initially, it was understood through the results of scenario building, where the analysis was carried out 'top down', and what was observed was how dangerous climate change was. This type of proposal is still widely used, especially when analysing global climate change and its consequences. However, the "top-down" analysis proposal does not highlight the local characteristics of adaptive capacity, so in order to remove the need to identify adaptive capacities on a smaller scale, a "bottom-up" systemic analysis was proposed, where the focus is on understanding how best to adapt to climate change (HUQ and REID, 2009).

Resilience is the fundamental basis of adaptation and the ability to adapt represents the pre-disposition to adaptation. Resilience thus represents the set of decision-making processes and the implementation of actions. In contrast, vulnerability is susceptible to changes in socio-environmental resilience. Reducing vulnerability therefore helps to boost a system's resilience (ADGER, 2004, 2006; EAKIE and WEHBE, 2009; FORSYTH, 2009).

In order to build resilience that strengthens socio-environmental systems, knowledge of the capacities of the system being analysed is crucial. It is in this sense that resilience emphasises the process of recovery, innovation and the capacity to learn and transform, focusing on future scenarios. Adaptive capacity is a fundamental part of vulnerability, which is one of the conditions before the event. Actions to deal with the impacts of climate change cannot be taken without first knowing the level of vulnerability of a socio-environmental system. Vulnerability is an inherent characteristic of every system, which is why it is important to establish acceptable levels of vulnerability, especially in highly vulnerable socio-environmental systems (CUTTER, 2008).

The study of vulnerability is still a very recent line of research, but it is important for climate change adaptation studies. Identifying the main factors that characterise the capacity of socio-environmental systems to respond to and recover from disasters is just as important as knowing the dynamics of the atmospheric systems that cause disasters. This capacity to adapt consists of identifying and strengthening the conditions inherent in socio-environmental systems that allow them to absorb impacts and deal with extreme events (ADGER, 2006; CUTTER, 2008; FOLKE, 2006).

To summarise, adaptive capacity refers to the potential of a social group or system to adapt in the context of climate change. A system that adjusts, modifies or alters itself with the aim of moderating potential damage or coping better with the consequences of a changing climate is considered to have a high adaptive capacity (ADGER, 2003; SMIT and WANDEL, 2006).

Climate variability is intrinsic to uncertainty and social groups are already facing and responding to this reality (ADGER, 2002). However, the way in which these social groups have applied the mechanisms seems to be inefficient in facing the new challenges brought about by climate change. The most vulnerable societies are not just those that experience the greatest impacts, but those that are more sensitive and have difficulties adapting to climate change and development pressures (VAN ASLST *et al.,* 2008).

The indicators selected by this research to make up the adaptive capacity sub-index represent four axes: education, income, infrastructure and public programmes for access to water. The indicators are: literacy rate; number of households earning 2 to 5 minimum wages; public investment in infrastructure and social programmes; and households with access to piped water.

Like the indicators in the previous sub-index, these indicators are also part of the CONDEPE/FIDEM municipal profile database. With the exception of the 'literacy rate' indicator, this indicator is the result of the 2010 census. However, it was obtained from the same database as the other indicators. The results of the indicators can be seen below:

Table 15: indicator results for each corresponding municipality.

MUNICIPALITIES	LITERACY	HOUSEHOLD	PUBLIC INVESTMENT	ADOMICILES

	RATE	NUMBERS WITH 2 A 5 WAGES	IN SOCIAL PROGRAMMES AND INFRASTRUCTURE	WITH ACCESS TO PIPED WATER
Arcoverde	61,96%	12,22%	11,30 %	49,62%
Buíque	53,33%	5,69%	13,20 %	35,47%
Custody	61,21%	12,81%	19,88 %	37,50%
Ibimirim	45,61%	9,02%	20,38 %	48,51%
Inajá	42,06%	7,05%	23,19 %	40,06%
Iguaraci	50,84%	9,92%	18,96 %	43,20%
Jatoba	55,79%	12,23%	0,00%	44,32%
Manari	41,75%	7,47%	16,47 %	45,54%
Sertânia	51,09%	10,73%	24,95 %	47,71%
Tupanatinga	40,27%	6,00%	15,15%	42,55%
Tacaratu	40,04%	8,84%	21,43%	43,02%

Source: prepared by the author.

Specifically, the values for the indicators "literacy rate" and "households with access to piped water" were positive, compared to the other two indicators. Low access to income and low public investment in social programmes and infrastructure reflect the fragility of these indicators in this context. The income indicator demonstrates what the poverty incidence indicator had already shown, which is that poverty is still a serious problem in the municipalities of the basin. Despite the existence of state policies to encourage the fight against climate change, the actual actions are still insufficient (Graph 25).

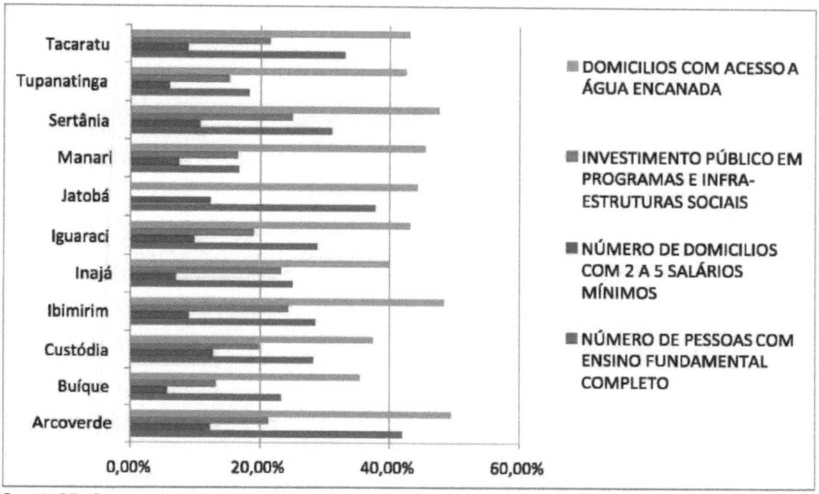

Graph 25: Graph of the results of the indicators of the adaptive capacity sub-index.
Source: prepared by the author.

To obtain the results of the other sub-indices, the values of the indicators of the adaptive capacity sub-index were applied to the equation: **Ip=Iobs-minimoI/maximoI-minimoIand** the final result was as follows:

Table 16: Results of the adaptive capacity sub-index.

MUNICIPALITIES	ADAPTATION SUB-INDEX
Arcoverde	98,3%
Buíque	86,0%
Custody	84,4%
Ibimirim	95,9%
Inajá	87,7%
Iguaraci	90,0%
Jatoba	72,5%
Manari	87,4%
Sertânia	100%
Tupanatinga	83,9%
Tacaratu	83,9%

Source: prepared by the author.

In order to rank the results of the sub-index, the same measure applied to the other sub-indices was adopted, following a standard that can be compared with each other (table 17):

Table 17: criteria for the severity of climate sensitivity in the basin's municipalities.

Adaptive capacity sub-index	0-40	41 - 60	61 - 100
Severity of adaptive capacity	Low	Average	High

Source: prepared by the author.

Even though two indicators had low values in all the municipalities, the final result of the sub-index was positive in all the municipalities in the basin (Figure 23).

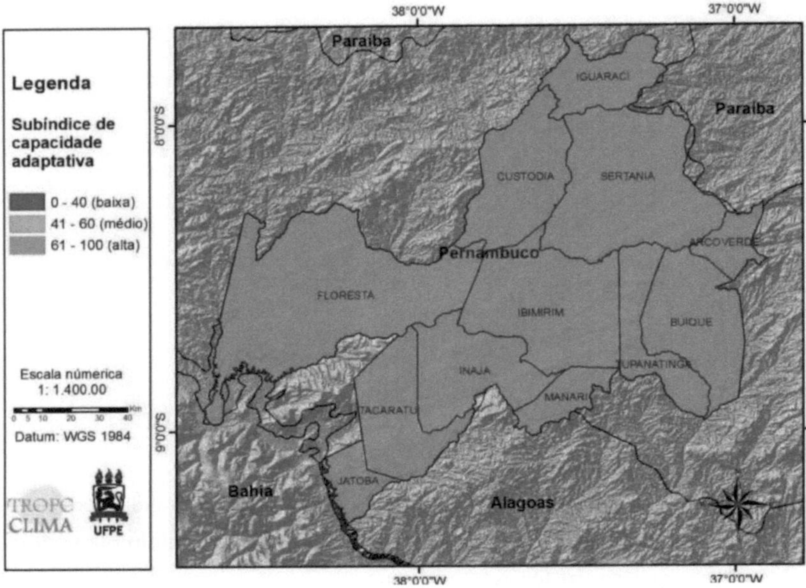

Figure 23: Map of the adaptive capacity sub-index. Source: prepared by the author.

The indicators that make up this sub-index represent four axes: human capital, family economic capital, public financial capital and natural capital. Each of these axes was understood according to attributes of the socio-environmental system. In this case, the Moxotó basin and its municipalities. Attributes of the population (human capital), their forms of organisation (family economic capital), economic variables (public financial capital) and access to natural resources (natural capital), were the attributes that determined the selection of these indicators to comprise the adaptive capacity sub-index.

All the municipalities analysed showed results above 80% in these sub-indices. The percentage of households with access to piped water represents a very favourable characteristic . Only the municipalities of Buíque and Custódia have less than 40% of households without access to piped water. However, this is not the ideal result for this indicator and expanding the structure of access to water for all is fundamental to mitigating the impacts of climate change.

Literacy programmes have boosted the literacy rate, especially among children and young people in the semi-arid region. However, despite the existence of specific programmes for youth and adult literacy, the percentage of literate people in this age group is still not what was expected (BARBOSA, *et al.,* 2013). However, the final result of the indicator is positive, considering that these percentages are from 2010 and that positive progress may already have been made.

With regard to the income indicator, the result shows that poverty is still a real problem in these municipalities, of which only five have more than 10 per cent of households with incomes between 2 and 5 minimum wages. All the others are below 10 per cent.

In the indicator of public investment in social programmes and infrastructure, results were obtained for all the municipalities, with the exception of the municipality of Jatobá, which, according to CONDEPE/FIDEM, the municipality's administration did not inform how much of its public finances were invested in social programmes and infrastructure.

The final result of the adaptive capacity sub-index indicates that the indicators have positive values, but that they need to be adjusted and provide elements for future adaptation plans.

3.2.4. index of socio-environmental vulnerability to climate change

Research into climate change has spread throughout the natural sciences, with the emphasis mainly on modelling future scenarios, focusing on understanding the impacts of global climate change (SIMON, 2007). However, recognising climate change as a problem whose character is based on epistemology as well as reality has gained ground in research, even in the latest IPCC report (LAMPIS, 2013).

Vulnerability to climate change can be understood from two paradigms. Firstly, as a fact related mainly to physical systems and, secondly, as a problem of the development of human systems. However, the real need for more concrete responses to the problem of climate change has led to the construction of the third paradigm, which consists of a combination of the two previous ones . This third paradigm has an integrated approach and is represented by three main approaches: 1. risk - threat; 2. social construction of risk; 3. integrated approach (LAMPIS, 2013; CUTTER *et al.,* 2003; ADGER, 1999).

Vulnerability to climate change is more than just exposure to some kind of risk. It occurs when the relationship between exposure to a risk generates vulnerability and consists of the resources, access and capital needed to face the risks without losing capacities and opportunities. When we talk about vulnerability to climate change, we are referring mainly to socio-environmental vulnerability, which is aggravated by the problems of limited human development (LAMPIS, 2010).

Basically, measuring vulnerability to climate change depends on analysing the effects of regional or local climate change and society's capacity to deal with these changes. Vulnerability to climate change is inherent to a given social system and varies according to levels of socio-economic development.

The spatio-temporal variability of rainfall in the semi-arid region does not guarantee the success of subsistence crops. According to Marengo (2008), frequent periods of drought represent major losses for the region. Also according to Marengo (2008), the Real-Time Monitoring Programme for the Northeast Region (Proclima) offers, among its products, a report on the number of water deficits during the rainy season and the author considers this product to be an indicator of the vulnerability of agriculture to the lack of rainfall in the region. However, to date there has been no scientific academic product or government report on quantifying and spatialising vulnerability to climate change in the semi-arid region.

In this context, the innovative nature of this research stands out, as it applies and adapts a methodology never before developed for a river basin in the semi-arid region. In addition to the objectives inherent in this research, these results seek to contribute indirectly to regional development, in the sense of providing useful information for future specific mitigation plans to deal with climate change.

The estimation of vulnerability to climate change proposed in this research was applied to the municipalities that make up the Moxotó river basin, considering indicators that reflect the degree of exposure to climate change, indicators of sensitivity to these changes as well as indicators that demonstrate the magnitude of adaptive capacity. The aim was to apply an integrated assessment of vulnerability, in which the indicators selected were representative of the actual biophysical conditions of the Moxotó basin and also described the socio-economic conditions of the municipalities in the basin.

The climate change vulnerability index is the result of the equation: **[Vulnerability = (exposure + sensitivity) - adaptive capacity].** The discussion of the results corresponding to each of the sub-indices was presented in the previous topics. However, for a better understanding of the final vulnerability result, we can look at the results of each sub-index in the table below:

Table 18: Results of the exposure, sensitivity and adaptive capacity sub-indices.

MUNICIPALITIES	**EXPOSURE SUB-INDEX**	**SUBINDICE** OF **SENSITIVITY**	**ADAPTATION SUB-INDEX**
Arcoverde	64%	63%	98,3%
Buíque	62%	84%	86,0%

Custody	22%	92%	84,4%
Ibimirim	65%	100%	95,9%
Inajá	33%	91%	87,7%
Iguaraci	-3%	97%	90,0%
Jatoba	56%	94%	72,5%
Manari	49%	96%	87,4%
Sertânia	51%	94%	100%
Tupanatinga	67%	95%	83,9%
Tacaratu	30%	100%	83,9%

Source: prepared by the author.

It can be seen that the result of the exposure sub-index for the municipality of Iguaraci was -3%. This is because this municipality had the lowest value in the indicators that make up this sub-index, and when these values were applied to the equation corresponding to the processing of the sub-indices for this municipality, the result was negative. However, it is clear that this municipality has low exposure, not negative exposure (Graph 26).

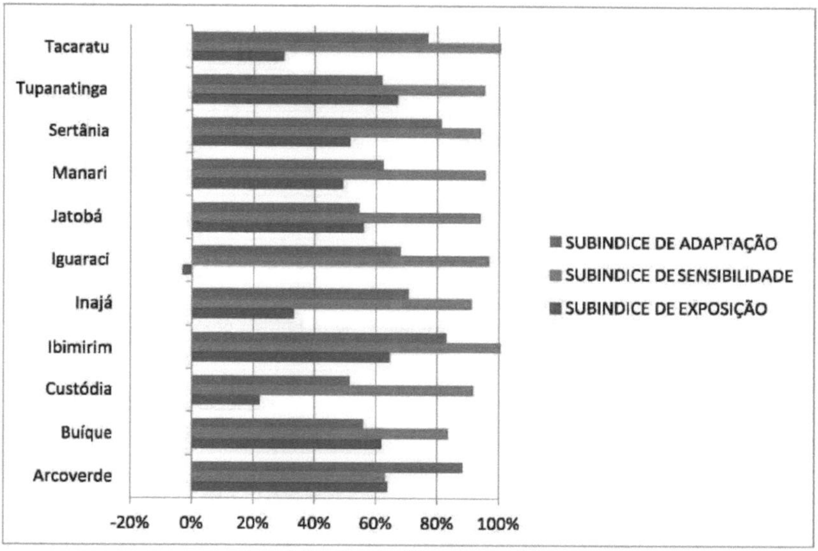

Graph 26: Graph of the results of the sub-indices.

Source: prepared by the author.

The results of each of the sub-indices were combined and applied to the vulnerability equation, where the final result is the climate change vulnerability index (Table 19 and Graph 27)

Table 19: Results of the exposure, sensitivity and adaptive capacity sub-indices.

MUNICIPALITIES	VULNERABILITY

Arcoverde	29%
Buíque	59%
Custody	30%
Ibimirim	69%
Inajá	37%
Iguaraci	4%
Jatoba	78%
Manari	57%
Sertânia	44%
Tupanatinga	78%
Tacaratu	47%

Source: prepared by the author.

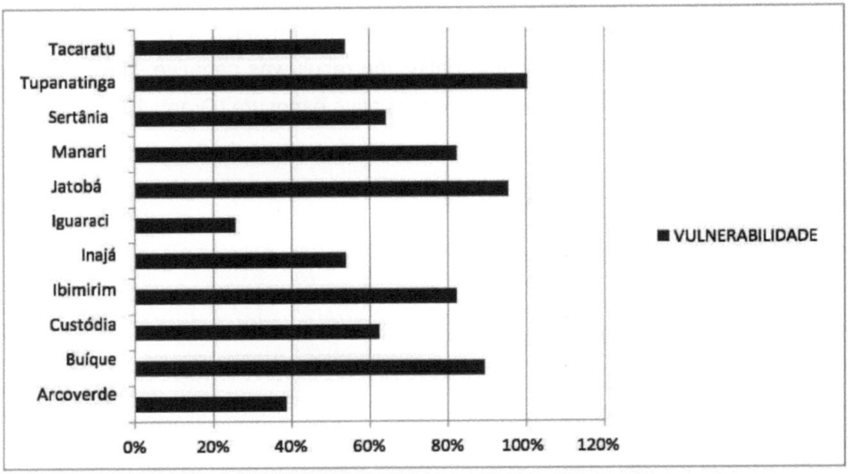

Graph 27: Graph of the results of the climate change vulnerability index. Source: prepared by the author.

As with all the other sub-index results, in the final index result the values were prioritised and spatialised in order to better compare the results (Table 20 and Figure 24).

Table 20: Criteria for the severity of climate change vulnerability index.

Climate change vulnerability index	0-40	41 -60	61-100
Severity of climate change vulnerability index	Low	Average	High

Source: prepared by the author.

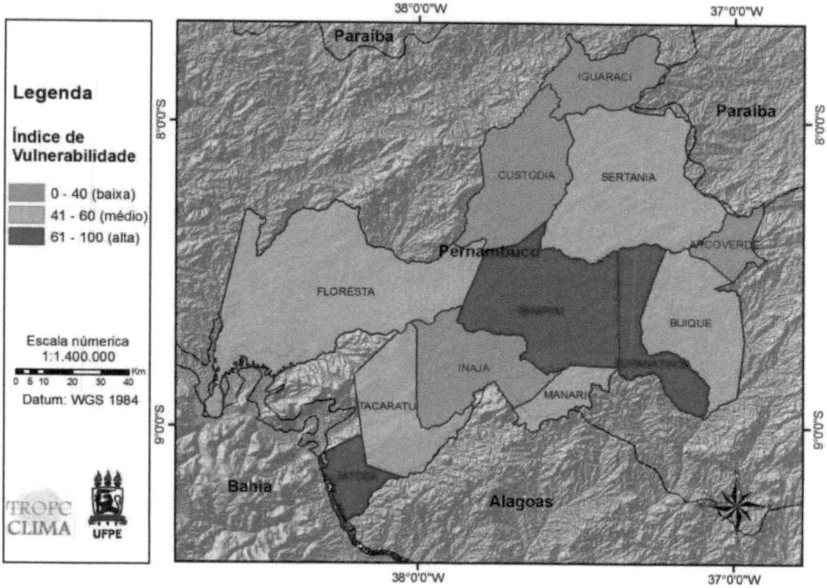

Figure 24: Map of the socio-environmental vulnerability index to climate change.
Source: prepared by the author.

The result of the index indicates that although the Moxotó basin is, according to the climate change perspective of global numerical models, under the same climate dynamics, the results of this research show the opposite. The smaller scale of the precipitation trend varies both spatially and temporally, as do the impacts that these changes produce. The effects of climate change in the basin area are not homogeneous, and therefore neither is the result of vulnerability.

The 12 municipalities in the basin that were analysed had varied results: 4 municipalities had low severity (0 - 40), 5 municipalities had medium severity (41 - 60) and only 3 municipalities had high severity (61 - 100).

The municipalities that showed the best results were Arcoverde, Custódia, Iguaraci and Inajá (0 - 40), with Iguaraci (4%) showing the lowest vulnerability value, followed by Arcoverde (29%), Custódia (30%) and Inajá (37%). Three of the municipalities in this group had high severity values in the exposure and sensitivity sub-indexes (Arcoverde, Custódia and Inajá), but also had the best results in the adaptive capacity sub-index, which resulted in positive vulnerability values. The exception is the municipality of Iguaraci, which had low exposure severity, a factor that favoured the municipality's low vulnerability score.

The second group is made up of municipalities with medium severity: Buíque, Manari, Sertânia and Tacaratu (41 - 60). Buíque (59%) was the municipality in this group with the highest index value, followed by Manari (57%). The other municipalities (Sertânia 44% and Tacaratu 47%) had similar values.

The third and final group represents the municipalities with high severity (61 - 100): Ibimirim, Jatobá and Tupanatinga. These municipalities are the most vulnerable in the Moxotó basin, with Jatobá (78%) and Tupanatinga (78%) having the highest values in this group, followed by Ibimirim (69%)

CHAPTER 4

FINAL CONSIDERATIONS

Based on what has been presented in the chapters that structure this book, it is possible to make the following final considerations.

Climatology in Brazil gradually adopted a more dynamic perspective in its studies, gradually abandoning the generalist view that was so common in the first climate studies carried out in the country. This dynamic perspective was based on a systemic view of the atmosphere-surface relationship. The influence of general systems theory not only brought dynamism to climate studies, but also brought them closer to other spheres of scientific study (ALVES, 2012).

Throughout the construction of this book, there has been a concern to address the general theory of systems and how it has influenced and influences geographical studies. The systemic theory that serves as the theoretical basis for much research in geography is one of the examples of this strong relationship between geography and the systemic vision.

Based on a systemic perspective, this research focused on investigating and analysing the relationship between the climate (and its changes and variability) and human activities, as well as the relationship between them in different forms in space.

The methodology adopted in this study is a reflection of how the systemic approach can be useful in scientific research, especially when it comes to analysing different types of systems in order to generate a single final result. The methodology used to develop the climate change vulnerability index is made up of physical, social and economic data. When different interconnected systems are analysed together, the final result is the spatialisation of the socio-environmental vulnerability of the Moxotó river basin.

The study area, the Moxotó basin, is located in the semi-arid region of north-eastern Brazil. This is a region that historically suffers periods of drought. Irregular rainfall and high climatic variability are some of the many factors that favour drought in this region.

Since 2012, the semi-arid region has been experiencing drought, which has resulted in economic losses for the region's municipalities, most of which depend on agricultural activities, leading to worsening situations of vulnerability social and environmental.

With this scenario of changes, climate variability and vulnerabilities in a region where poverty is still very present, there is a need to add 'new' spheres of analysis to climate change research, such as reducing vulnerability, increasing resilience and adaptive capacity.

Prior to the quantitative and qualitative definition of socio-environmental vulnerability in the Moxotó basin, the dynamics of rainfall over the last 84 years were analysed, as was the behaviour of the vegetation over the last 27 years.

When analysing the dynamics of the results, the east, centre and south sectors showed a significant positive trend. The western sector had a significant negative trend and the northern sector showed a non-significant trend. A temporal concentration of precipitation was also observed in some months of the year. This result was also presented by sector: 66% of precipitation in the eastern sector is concentrated in the months of March, April, May, June and July; 51% of precipitation in the western sector is concentrated in the months of February, March and April; the central sector concentrates 67% of precipitation in the months of January, February, March and April; the southern sector concentrates 66% of precipitation in the same months as the central sector, and the northern sector concentrates 68% of precipitation in the aforementioned months.

Thus, the profile of rainfall trends analysed during the period from 1930 to 2014 shows spatial irregularity and temporal concentration during the months of January, February, March and April. The results related to the spatio-temporal analysis of vegetation cover using NDVI showed three categories of vegetation according to vegetation density. It was observed that there was no relationship between the sectors of the basin that showed a positive trend in rainfall and the sectors with the highest vegetation density. Topography has a strong influence on the basin, and one of the results of this stage is the concentration of the basin's native vegetation in the higher areas.

It was also possible to identify the presence of this vegetation in all the images analysed in this series. It was observed that of the three categories of vegetation density present, the most abundant is category 1, which represents low-density vegetation (NDVI value 0.25 - 0.50), and the least abundant is category 3, which represents high-density vegetation (NDVI value 0.75-1).

It was found that vegetation grew as follows according to density category: density category 1 (0.25 - 0.50) had a growth rate of 8.15 per cent; density category 2 (0.50 - 0.75) had a growth rate of 16.5 per cent, making it the category that grew the most in the basin area; and lastly, density category 3 (0.75 - 1) had a growth rate of 3.38 per cent. It was also possible to generate the growth rate of exposed soil as a secondary result. This result was used as an indicator to integrate the exposure sub-index.

The results of the chapters on analysing rainfall trends and the spatio-temporal analysis of vegetation cover served as a risk model, in which there is a description of recent impacts on the basin's landscape, where it was possible to carry out a preliminary diagnosis of its vulnerable characteristics before drawing up the index.

The data generated in both chapters was also applied as indicators in the modelling of the climate change vulnerability index. It was concluded that even though there are more sectors with a positive rainfall trend, the rainfall recorded during the series analysed is concentrated in a few months, which generates water stress in the other months of the year and, consequently, causes impacts on the vegetation.

Even before putting the index into practice, some municipalities in the basin were visited with the aim of reconnoitring the study area, establishing dialogues with some residents and making a preliminary diagnosis of the landscape. The conclusion of these visits was that, even though they were only reconnaissance visits, what was recorded through images already indicated a situation susceptible to vulnerability. It is important to emphasise that all the visits took place during 2012 and 2013, a period in which the semi-arid region was facing the onset of a drought.

The results of the index have been presented separately according to the results of each indicator and corresponding sub-index. The sub-indices are divided into three: exposure, sensitivity and adaptive capacity.

The exposure sub-index was made up of physical indicators: rainfall trends, rate of exposed soil and temperature variation. It was concluded that, according to the results of this sub-index, four municipalities had high exposure (Arcoverde, Buíque, Ibimirim and Tacaratu); four had medium exposure (Jatobá, Manari and Sertania); and another four had low exposure (Custódia, Iguaraci, Inaja and Tupanatinga). These results refer to the level of exposure to climate change of each of these municipalities: the greater the exposure, the greater the susceptibility to impacts.

However, the exposure results are insufficient to state how vulnerable these municipalities are, since the results only conclude the level of exposure from the point of view of the physical factors of the landscape. In order to generate the final result of the index, it was necessary to analyse the sensitivity of the social groups that occupy the municipalities as well as the adaptive capacity that these groups have to face the adversities of climate change.

The sensitivity sub-index is basically linked to the concepts of stability, resilience and flexibility. It refers to the ability of a given system (physical or social) to withstand the impacts of climatic disturbances, in such a way that the impacts generated by the disturbances are minimal. To make up this sub-index, the indicators selected were: number of workers in primary activities; number of private households without basic sanitation;

number of public health establishments and incidence of poverty.

The results of this sub-index were quite negative. Only the municipality of Arcoverde showed a medium sensitivity result. All the others showed high sensitivity, which indicates that these municipalities do not have enough stability or resilience to cope with the disturbances caused by climate change.

The adaptive capacity sub-index represents the human capital and social infrastructure existing in the municipalities that allow them to provide real conditions for adapting to future changes. This sub-index is made up of the following indicators: literacy rate; number of households earning 2 to 5 (or more) minimum wages; public investment in social programmes, infrastructure and households with access to piped water.

The weight of each indicator in the construction of this sub-index was distributed equally; however, it is important to emphasise that thanks to the positive values of the literacy rate and households with access to piped water indicators, the final result of the adaptive capacity sub-index is high for all municipalities. However, the indicators for the number of households earning 2 or 5 (or more) minimum wages and public investment in social programmes and infrastructure show that there is still a deficiency in income distribution in these municipalities. Only the municipalities of Arcoverde and Custódio have more than 12 per cent of households earning 2 to 5 (or more) minimum wages. Even so, the percentages of public investment in infrastructure and social programmes are quite low.

A simple algebra equation was applied to generate the final index result. The final results of each of the sub-indices described above are derived from the vulnerability, which would be the result of adding the results of the exposure and sensitivity sub-indices minus the result of the adaptive capacity sub-index.

The results indicate that there is no homogeneity of vulnerability in the basin area. Thanks to the methodology applied on a smaller scale, it is possible to observe variations in variability that could not be observed if they had been applied on another scale such as global models, even if applied to *downscaling.*

The results were that Arcoverde, Custódia, Iguaraci and Inajá showed low vulnerability; Buíque, Manari and Tacaratu showed medium vulnerability; and the most critical municipalities were Jatobá, Ibimirim and Tupanatinga, with high levels of vulnerability to climate change.

The conclusion is that this is a methodology that responds well when applied on a municipal scale, that the watershed cut-off represented the best option for observing changes in climate and vegetation, and that, especially in the rural areas where the study was carried out, urbanisation is not yet as impactful as in the capital cities, so the municipal cut-off would not be as useful for this type of analysis as the watershed cut-off, which has water dividers and influences the distribution of rainfall.

This study is important as a strategic product for planning adaptation to climate change and for setting targets to minimise actions that contribute to landscape impacts and climate change. However, this study alone will not be enough to promote sustainable development that will boost resilience in the Moxotó river basin. It is therefore suggested that studies related to climate change research be integrated with public policies that boost the resilience and adaptive capacity of areas such as the Moxotó catchment, which has high levels of vulnerability.

CHAPTER 5

BIBLIOGRAPHICAL REFERENCES

ADGER, W. N. Social vulnerability to climate change and extremes in Coastal Vietnam. **World Development,** v. 27, n. 2, p. 249-269, 1999.

ADGER, W. N. Adaptation to climate change in the developing world. **Progress in Development Studies,** v. 3, p. 179-195, 2003.

ADGER, W. N. The right to keep cold. **Environment and planning A,** v. 36, p. 1711- 1715, 2004.

ADGER, W. N. Vulnerability. **Global Environmental Change,** v. 16, p. 268-281, 2006.

ADGER, W. N. et al. Adapatation to climate change in the developing world. **Progress in Development Studies,** v. 3, p. 179-195, 2002.

AHLSTRÕM, A. et al. The dominant role of semi-aid ecosystems in the trend and variability of land CO2 sink. **Science,** v. 348, p. 1-19, May 2015.

ALMEIDA, H. A. **Probability of rainfall in the southeast of Bahia. Technical Bulletin No. 182.** CEPLAC\CEPEC. Ilhéus, p. 32. 2001.

ALVES, J. M. B.; REPELLI, C. A. Rainfall variability in the northern sector of the Northeast and the El Nino Southern Oscillation events. **RBMet,** 1992.

ALVES, K. M. A. S. **Profetas do Sertão:** um estudo sobre as previsões de sertanejos que se guiam pela natureza para enfrentar a seca. Sertões: Spaces, Times, Movements. Recife: [s.n.]. 2006.

ALVES, K. M. A. S. et al. **Case study of dynamic aspects of a Mesoscale Convective Complex (MCC).** XV Brazilian Symposium on Remote Sensing. Curitiba: [s.n.]. 2011.

ANA. Special section on the water crisis - special report 2014.
National Water Agency. Brasilia. 2014.

ASA. http://www.asabrasil.org.br/noticias, 12 December 2013. Accessed on: 10 June 2014.

AYOADE, J. O. **Introduction to climatology for the tropics.** Rio de Janeiro: Bertrand, 1991.

BARBOSA, M. R.; LOPES, M. A. C.; ROAZZI, A. A. The literacy of young people and adults in the Xingó programme: A look at its former students three years later. **Lugares de Educação journal,** v. 3, n. 6, p. 137-151, jul-dez 2013.

BATISTA, A. A. G. **Literacy, reading and writing.** Ministry of Education. Brasília, p. 13-17. 2006.

BAWDEN, D. Revisión de los conceptos de alfabetización informacional y alfabetización digital. **Anales de documentación,** v. 5, p. 361-408, 2002.

BIRKMANN, J.; NISHARA, F.; HETTIGE, S. Measuring vulnerability in Sri Lanka at the local levei. In: BIRKMANN, J. **Measuring vulnerability to natural hazards:** towards disaster resilient societies. Tokyo, New York. Paris: United Nations - University Press, 2006. Chap. 18, p. 329.

BLAIKIE, P. et al. **Vulrabilidad: el entorno social, político y económico de los desastres.** Red de Estudios Sociales en Prevención de Desastres en América Latina. Lima. 1996.

BROOKS, N.; ADGER, W. N.; KELLY, P. M. The determinais of vulnerability and adaptive capacity at the national levei and the implications for adaptation. **Global Environmental Change,** v. 15, p. 151-163, 2005.

BUAINAIN, A. M.; GARCIA, J. R. Rural development in the Brazilian semi-arid region: recent transformations, challenges and perspectives. **Confins,** v. 19, November 2013.

CASTRO, A. L. C. **Disaster manual.** Ministry of National Integration. Brasília. 2003.

CENDRERO, A. **Riesgos naturales y impacto ambiental.** Madrid: University-Business Foundation, 1997.

CHRISTOFOLETTI, A. **Modelling Environmental Systems.** São Paulo: Blucher, 1999.

COELHO, M. D. S.; GAN, M. A.; CONFORTE, J. C. Study of the variability of the position and cloudiness associated with the Atlantic ITCZ during the rainy season of 1998 and 1999 in the Brazilian Northeast. **RBMet,** 2004.

CONDEPE\FIDEM. **Municipal profile of the state of Pernambuco.** Pernambuco State Planning and Research Agency. Recife. 2010.

CONNOR, R. F.; HIROKI, K. Development of a method for assessing floyd vulnerablity. **WaterScience and Technology,** v. 51, n. 5, p. 61-67, 2005.

CPRM. **Geoenvironmental database of the Moxotó river basin -** PE. Hydrogeology Programme of Brazil. [S.I.J. 2004.

CUTTER, S. L. Vulnerability to environmental hazerd. **Progress in Human Geography,** v. 20, p. 529-539, 1996.

CUTTER, S. L. The science of vulnerability: models, methods and indicators. **Revista Crítica de Ciências Sociais,** v. 93, n. 1, p. 1-12, 2011.

CUTTER, S. L. et al. Social vulnerability to environmental hazards. **Social Science Quarterly**, v. 84, n. 2, p. 242-261,2003.

CUTTER, S.; FINCH, C. Temporal and spatial changes in social vulnerability to natural hazards. **Proceeding of the National Academy of Sciences** , v. 7, n. 105, p. 2301-2306, 2008.

DEMATTÊ, J. A. M.; TOLEDO, A. M. A.; SIMÕES, M. S. Methodology for recognising three soils using sensors: Laboratory and orbital. **Revista brasileira de ciência do solo,** v. 28, n. 1, p. 877-889, 2004.

EAKIN, H. C.; WEHBE, M. B. Linking local vulnerabulity to system sustainability in a resilience framework: two houses from Latin America. **Climate Change,** v. 93, p. 355- 377, 2009.

ESPÍNDOLA, G. M.; SANTOS, M. S.; ANDRADE, M. A. **The incidence of poverty in Brazil:** An empirical analysis, 1992-2005. National conference on public policies against poverty and inequality. Natal: [s.n.]. 2010.

FOLKE, C. Resilience: the emergence of a perspective for social-ecologiacal system analyses. **Global Environmental Change,** v. 16, p. 253-267, 2006.

FORSYTH, T. Community-Based adaptation to climate change.
Environment: Science and Policy for Sustainable, v. 51, n. 4, p. 22-31, August 2009.

FOWLER, H. J.; KILSBY, C. G.; 0'CONNELL, P. E. Modelling the impacts of climatic change and variability on the reliabity, resilience, and vulnerability of a water resource system. **Water Resources Research,** v. 39, n. 2, p. 1222, 2003.

HUQ, S.; REID, H. A. **A vidal approach to the threat climate change poses to the poor.** London: IIED, 2009.

IBGE. **Statistical data report on material losses during the 2012 drought.** Brazilian Institute of Geography and Statistics. Brasilia. 2012.

IPCC. **Managing the risks of extreme weather events and disasters to improve adaptation to climate change.** Group
Intergovernmental Panel of Experts on Climate Change. [S.l.], p. 16-32. 2012.

IPCC, I. P. O. C. C. **Climate Change 2001: The scientific basis. Contribution of Working Group I to the Third Assessment Report of the Intergovernmental Panei on Climate Change.** Cambridge University Press. [S.l.], p. 881. 2001.

IPCC, P. I. S. M. C. **5th Assessment Report of the Intergovernmental Panel on Climate Change: Impacts, Adaptation and Vulnerability.**

Cambridge University Press. [S.I.]. 2014.

IPEA. **Altlas da Vulnerabilidade Social nos Municípios Brasileiros.** Institute for Applied Economic Research. Brasília, p. 1-84. 2015.

KOUSKY, V. E.; GAN, M. A. Upper tropospheric cyclonic vortices in the tropical South Atlantic. **Academic Press,** Tellus, p. 538-551, 1981.

LAING, A. G.; FRITSCH, J. M. The global population of mesoscale convective complexes. **Quart. j. Roy. Meteor. Soc.,** v. 123, p. 389-405, 2006.

LAMPIS, A. **Pobreza y riesgo ambiental:** un problema de vulnerabilidad y desarrollo. Bogotá: Interdisciplinary Centre for Development Studies, 2010. Accessed on: 10 December 2013.

LAMPIS, A. **Adaptation to climate change:** the challenge of dual agendas. Climate change, social movements and public policies: a necessary link. Santiago de Chile: CLACSO, 2013.

LIMA, W. P.; ZAKIA, M. J. B. Hydrology of riparian forests. In: RODRIGUES, R. R.; LEITÃO FILHO, H. F. **Mata ciliares:** conservação e recuperação. São Paulo: University of São Paulo, 2000. p. 33-43.

LUERS, A. L. et al. A method for quantifying vulnerability, applied to the agricultural system of the Yaqui Valley, Mexico. **Global Environmental Change,** v. 13, p. 255-267, 2003.

MARANDOLA JR, E. Tangenciando a vulnerabilidade. In: HOGAN, D. M. J. E. **Population and climate change:** human dimensions of global environmental change. Campinas: NEPO - Unicamp, 2009.

MARANDOLA JR., E.; MELLO, L. F. D. **"Place" and "living space":** new approaches to planning and participation? Latin American Meeting of Geographers - Proceedings and Scientific Contributions - USP. São Paulo: [s.n.]. 2005.

MARENGO, J. A. **Global climate change and its effects on biodiversity - characterisation of the current climate and climate definitions for the Brazilian territory throughout the 20th century** Ministry of the Environment. Brasilia. 2007.

MARENGO, J. A. **Vulnerability, Impacts and Adaptation to Climate Change in the Brazilian Semi-Arid.** Strategic Partnerships/Strategic Studies Management Centre - Ministry of Science and Technology. Brasília, p. 149-176. 2009.

MARENGO, J. A. et al. **Characterisation of the current climate and definition of climate change for the Brazilian territory throughout the 20th century.** CPTECMNPE and IAG\USP. São Paulo. 2007.

MARENGO, J. A. et al. The Dorught of Amazonia in 2005. **Journal Climate,** n. 21, p. 495-516, 2008.

MARENGO, J. A. et al. Climate variability and change in the Brazilian semi-arid region. **Strategic Partnerships,** v. 27, p. 385-422, 2008.

MARENGO, J. A.; LACERDA, F. F. **Water resources in arid and semi-arid regions.** National Semi-Arid Institute. Campina Gramde , p. 385-422. 2011.

MARENGO, J. et al. Extreme Seasonal Events: Droughts from 2005 to 2010 and Floods in 2009. In: BETTS, R. (. **Risks of Climate Change in Brazil.** São José dos Campos: INPE/MOHC, 2011. p. 55.

MASSON, V. E. A. Holocene climate variability in Antarctica based on 11 ice- core isotopic records. **Quaternary Research,** v. 54, n. 3, p. 348-358, 2000.

MCCARTHY, J. J. et al. **Climate change 2001:** Impacts, adaptation and vulnerability. Cambridge: Cambridg University Press, 2001.

MCGARRY, **K. The dynamic context of information:** an introductory analysis. Brasília: Brinquet de Lemos, 1999.

MELO, C. R. **Análise do eixo leste da transposição do rio São Francisco fase aos cenários de uso previstos (Master's thesis).** Recife: UFPE, 2010.

MENDONÇA, M.; ROMERO, H. Natural hazards and subjective evaluation in the construction of social vulnerability to natural disasters in Chile and Brazil. **Inter. Interdisc. INTERthesis,** Florianópolis, v. 9, n. 1, p. 127-180, jan-jul 2012.

MENDONÇAS, F.; SOUZA, A. V.; DUTRA, D. A. Saúde pública, urbanização e dengue no Brasil. **Sociedade e Natureza,** Uberlândia, v. 21, n. 3, p. 257-269, December 2009.

MENDOZA, G. A. et al. **The criteria and indicators toolbox series guidelines for applying multi-criteria analysis to the assessment of criteria and indicators.** Centre for International Forestry Research (CIFOR). [S.l.]. 1999.

MONTERROSO, A. I. R. **Contribuición al estúdio de la vulnerabilidad al cambio climático en México (Doctoral Thesis).** Ciudad de México: Postgraduate Programme in Geography - National Autonomous University of Mexico, 2012.

MOSSET, R.; BRENKER, A.; MALONE, E. **Measuring vulnerability: a trial indicator set.** Paficic Norwest Laboratories. [S.l.], p. http://www.pnl.gov/globalchange/projects/vul/indicators.pdf. 2000.

NOBREGA, R. S. **The El Nino phenomenon and the joint influence of the Altantic dipole in the state of Ceára.** Brazilian Congress of Meteorology. Florianópolis: [s.n.]. 2000.

NORQUIST, D. C.; RECKER, E. E.; REED, R. J. The energetics of African wave disturbances as observed during Phase III of GATE. **Mon. Wa. Rev.,** p. 334-342, 1977.

OBERMAIER, M.; ROSA, L. P. Climate change and adaptation in Brazil: A critical analysis. **Estudos Avançados,** São Paulo, v. 27, n. 78, p. 155-176, 2013.

0'BRIEN, K. L. et al. Mapping vulnerability to multiple stressors: climate change and globalisation in India. **Global Environmental Change,** v. 14, p. 303-313, 2004.

OLIVEIRA, G. C. S.; ALVES, K. M. A. S.; NOBRÉGA, R. S. **Main mechanisms responsible for the occurrence of rain in the northeastern semi-arid region.** II Workshop on Climate Change and Water Resources in Pernambuco. Recife: [s.n.]. 2010.

OLIVEIRA, T. H. **Balanço de radiação da bacia hidrográfica do rio Moxotó - PE atrvés de imagens TM Lansat-5 e ETM+ Landsat-7 (monografia de conslução de curso).** Recife: Department of Geographical Sciences - Federal University of Pernambuco, 2009.

OLIVEIRA, T. H.; GALVÍNCIO, J. D. Environmental Characterisation of the Moxotó River Basin - PE Using Thermal Remote Sensing. **Revista brasileira de Geografia Física,** Recife, v. 1, n. 2, p. 30-49, May 2009.

UN. **Bali Action Plan.** United Nations National Conference on Climate Change. Indonesia. 2007.

PARMESAN, C. Ecological and evolutionary responses to recent climate change. **Annu. Rev. Ecol. Evol Syst,** v. 37, p. 637-669, 2006.

PERAZZO, G. M. **Profile of environmental sanitation in 29 municipalities in the Xingó area.** University Press. Recife. 2002.

UNDP. **Report on human development to eradicate poverty.** United Nations Human Development Programme. New York. 1997.

UNDP. **Human Development Report 2014 - Sustaining human progress: Reducing vulnerabilities and strengthening resilience.**
United Nations Development Programme. [S.l.J. 2014.

ROMERO, H. et al. **Changes in land use and land cover associated with the urbanisation of Chilean metropolises.** Anales Sociedad Chilena de Ciencias Geográficas. Santiago: [s.n.]. 2006.

ROMERO, H. et al. Urban climates and atmospheric pollution in Santiago de Chile. **EURE**

(Santiago), v. 36, n. 109, p. 35-62, Diciembre 2010.

SECTMA. **Water Demand and Supply in Pernambuco:** An Indicative Approach. Recife: Secretariat of Science, Technology and the Environment - SECTMA, 2009.

SEPLAN. **Report on the physical characteristics of the Moxotó River Basin.** Planning and Management Secretariat of the State of Pernambuco. Recife. 1993.

SIMON, D. Cities and global environmental change: Exploring the links. **The geographical journal,** v. 173, n. 1, p. 75-79, 2007.

SMIT, B.; WANDEL, J. Adaptation, adaptive capacity and vulnerability. **Global Environmental Change,** v. 16, p. 282-292, 2006.

SMITHERS, J.; SMIT, B. Human adaptation to climatic variability and change. In: SCHIPPER, E.; BURTON, I. **Adaptation to Climate Change.** London: Earthcan, 2009. p. 15-33.

SOUZA, R. C.; CAMARGO, M. E. **Time series analysis and forecasting:** the ARIMA models. Rio de Janeiro: Gráfica e editora Regional, 2004.

SULLIVAN, C.; MEIGH, J. Targeting attetion on local vulnerabilities using an integrated Index approach: the example of the climate vulnerability Index. **WaterScience and Technology,** v. 51, n. 5, p. 69-78, 2005.

TAVARES, B. A. C. A **participação da morfoestrutura na génese da compartimentação geomorfológica do gráben do Cariatá - Paraíba (Master's dissertation).** Recife: UFPE, 2010.

UNIVALE-DENOCS. **Statistical and historical documents.** UUNIVALE-DENOCS. Ibimirim. 2013.

VAN AALST, M. K.; CANNON, T.; BURTON, I. Community levei adaptation to climate change: the potential role of participatory community risk assesment. **Global Environmental Change,** v. 18, p. 165-179, 2008.

VASCONCELOS, P. B. **Interferências do sistema territorial sobre a formação da cooperação:** Indícios a partir do perímetro irrigado do Moxotó - PE (Master's thesis). Recife: Department of Geographical Sciences - Federal University of Pernambuco, 2009.

WISNER, B. et al. **At risk.** London: Routledge, 2004.

MIX
Papier aus verantwortungsvollen Quellen
Paper from responsible sources
FSC® C105338

Printed by Books on Demand GmbH, Norderstedt / Germany